METHODS IN MOLECULAR BIOLOGY

For further volumes:
http://www.springer.com/series/7651

For over 35 years, biological scientists have come to rely on the research protocols and methodologies in the critically acclaimed *Methods in Molecular Biology* series. The series was the first to introduce the step-by-step protocols approach that has become the standard in all biomedical protocol publishing. Each protocol is provided in readily-reproducible step-by-step fashion, opening with an introductory overview, a list of the materials and reagents needed to complete the experiment, and followed by a detailed procedure that is supported with a helpful notes section offering tips and tricks of the trade as well as troubleshooting advice. These hallmark features were introduced by series editor Dr. John Walker and constitute the key ingredient in each and every volume of the *Methods in Molecular Biology* series. Tested and trusted, comprehensive and reliable, all protocols from the series are indexed in PubMed.

Plant Iron Homeostasis

Methods and Protocols

Edited by

Jeeyon Jeong

Department of Biology, Amherst College, Amherst, MA, USA

Humana Press

Editor
Jeeyon Jeong
Department of Biology
Amherst College
Amherst, MA, USA

ISSN 1064-3745 ISSN 1940-6029 (electronic)
Methods in Molecular Biology
ISBN 978-1-0716-3182-9 ISBN 978-1-0716-3183-6 (eBook)
https://doi.org/10.1007/978-1-0716-3183-6

Cover Illustration Caption: Adapted from the RGB elemental map of iron (red), calcium (green), and zinc (blue) of an Arabidopsis thaliana stem obtained by Ju-Chen Chia, Louisa Smieska, Arther Woll and Olena Vatamaniuk using confocal synchrotron x-ray fluorescence microscopy (Chapter 14).

This Humana imprint is published by the registered company Springer Science+Business Media, LLC, part of Springer Nature.
The registered company address is: 1 New York Plaza, New York, NY 10004, U.S.A.

Preface

The nutritional status of iron directly impacts plant growth and development, as well as the yield and nutritional quality of crops. Iron is an essential micronutrient that serves as a cofactor in numerous metabolic processes, but excess iron is harmful. Due to extremely low availability of iron under aerobic physiological conditions, plants face an additional challenge to meet their iron demand while preventing iron-induced toxicity. Thus, plants have evolved delicate mechanisms to tightly control iron.

This volume of *Methods in Molecular Biology* focuses on iron homeostasis in plants. The specific contents of the chapters range from protocols to study the iron deficiency response, the interaction between root and microbes under iron-deficient conditions, the transcriptional network of iron homeostasis, systemic signaling of iron, chloroplast iron regulation, as well as methods on quantitative proteomics, histochemical iron staining, and metal imaging using x-ray fluorescence microscopy. In addition, a report of a comparative survey of iron homeostasis genes regulated by chromatin modifiers and a review article on iron retrograde signaling in plants are included. While most chapters are based on research with *Arabidopsis*, this volume also covers a protocol on a semi-hydroponic system to study iron homeostasis in maize, and an optimized method for cultivating algae for iron nutrition research. This book will serve as a valuable resource for the plant iron homeostasis research community, and will be of broad interest to plant biologists, soil scientists, and molecular biologists.

Amherst, MA, USA *Jeeyon Jeong*

Contents

Contributors

PETRA BAUER • *Institute of Botany, Heinrich Heine University, Düsseldorf, Germany; Cluster of Excellence on Plant Sciences (CEPLAS), Heinrich Heine University, Düsseldorf, Germany*

JU-CHEN CHIA • *Plant Biology Section, School of Integrative Plant Science, Cornell University, Ithaca, NY, USA*

GENEVIÈVE CONJERO • *IPSiM, Univ. Montpellier, CNRS, INRAE, Institut Agro, Montpellier, France*

DAYISHAA DAGA • *Department of Biology, Amherst College, Amherst, MA, USA*

XIAOYING DONG • *State Key Laboratory of Soil and Sustainable Agriculture, Institute of Soil Science, Chinese Academy of Sciences, Nanjing, China*

CHRISTIAN DUBOS • *IPSiM, Univ. Montpellier, CNRS, INRAE, Institut Agro, Montpellier, France; Campus INRAE, SupAgro, Institute for Plant Sciences of Montpellier (IPSiM), Montpellier, France*

STAVROULA FILI • *Department of Biology, University of Massachusetts Amherst, Amherst, MA, USA*

FEI GAO • *Campus INRAE, SupAgro, Institute for Plant Sciences of Montpellier (IPSiM), Montpellier, France*

ANNE G. GLAESENER • *California Institute for Quantitative Biosciences (QB3), University of California, Berkeley, CA, USA*

DIEGO F. GOMEZ-CASATI • *Centro de Estudios Fotosintéticos y Bioquímicos (CEFOBI-CONICET), Universidad Nacional de Rosario, Rosario, Argentina*

LOUIS GRILLET • *Department of Agricultural Chemistry, National Taiwan University, Taipei, Taiwan*

EN-JUNG HSIEH • *Department of Agricultural Chemistry, National Taiwan University, Taipei, Taiwan*

SHU-HUA HSU • *Plant-Microbe Interactions, Department of Biology, Utrecht University, Utrecht, the Netherlands*

JEEYON JEONG • *Department of Biology, Amherst College, Amherst, MA, USA; Program in Biochemistry and Biophysics, Amherst College, Amherst, MA, USA*

SUN A. KIM • *Department of Biological Sciences, Dartmouth College, Hanover, NH, USA*

PING LAN • *State Key Laboratory of Soil and Sustainable Agriculture, Institute of Soil Science, Chinese Academy of Sciences, Nanjing, China; University of Chinese Academy of Sciences, Beijing, China*

JOOHYUN LEE • *Division of Natural and Applied Sciences, Duke Kunshan University, Jiangsu, China*

CLAIRE MACERO • *Department of Biology, Amherst College, Amherst, MA, USA*

SABEEHA S. MERCHANT • *California Institute for Quantitative Biosciences (QB3), University of California, Berkeley, CA, USA; Department of Molecular and Cell Biology, University of California, Berkeley, CA, USA; Department of Plant and Microbial Biology, University of California, Berkeley, CA, USA; Lawrence Livermore National Laboratory, Physical and Life Science Directorate, Livermore, CA, USA*

JULIE NEVEU • *Plant Science Research Laboratory (LRSV), CNRS/University of Toulouse, Auzeville Tolosane, France*

MARY NJERI NGIGI • *Institute of Botany, Heinrich Heine University, Düsseldorf, Germany; Cluster of Excellence on Plant Sciences (CEPLAS), Heinrich Heine University, Düsseldorf, Germany*
SARA OMER • *Department of Biology, Amherst College, Amherst, MA, USA; Program in Biochemistry and Biophysics, Amherst College, Amherst, MA, USA*
MARIA A. PAGANI • *Centro de Estudios Fotosintéticos y Bioquímicos (CEFOBI-CONICET), Universidad Nacional de Rosario, Rosario, Argentina*
YILIN PAN • *State Key Laboratory of Soil and Sustainable Agriculture, Institute of Soil Science, Chinese Academy of Sciences, Nanjing, China; University of Chinese Academy of Sciences, Beijing, China*
CORNÉ M. J. PIETERSE • *Plant-Microbe Interactions, Department of Biology, Utrecht University, Utrecht, the Netherlands*
KEVIN ROBE • *IPSiM, Univ. Montpellier, CNRS, INRAE, Institut Agro, Montpellier, France*
HANNETZ ROSCHZTTARDTZ • *Departamento de Genética Molecular y Microbiología, Facultad de Ciencias Biológicas, Pontificia Universidad Católica de Chile, Santiago, Chile*
RENFANG SHEN • *State Key Laboratory of Soil and Sustainable Agriculture, Institute of Soil Science, Chinese Academy of Sciences, Nanjing, China*
LOUISA SMIESKA • *Cornell High Energy Synchrotron Source (CHESS), Cornell University, Ithaca, NY, USA*
ÁDÁM SOLTI • *Department of Plant Physiology and Molecular Plant Biology, Institute of Biology, ELTE Eötvös Loránd University, Budapest, Hungary*
JULIEN SPIELMANN • *Plant Science Research Laboratory (LRSV), CNRS/University of Toulouse, Auzeville Tolosane, France*
MAX J. J. STASSEN • *Plant-Microbe Interactions, Department of Biology, Utrecht University, Utrecht, the Netherlands*
IOANNIS A. STRINGLIS • *Plant-Microbe Interactions, Department of Biology, Utrecht University, Utrecht, the Netherlands*
JOAQUÍN VARGAS • *Departamento de Genética Molecular y Microbiología, Facultad de Ciencias Biológicas, Pontificia Universidad Católica de Chile, Santiago, Chile*
OLENA K. VATAMANIUK • *Plant Biology Section, School of Integrative Plant Science, Cornell University, Ithaca, NY, USA*
GRÉGORY VERT • *Plant Science Research Laboratory (LRSV), CNRS/University of Toulouse, Auzeville Tolosane, France*
ELSBETH WALKER • *Department of Biology, University of Massachusetts Amherst, Amherst, MA, USA*
RUONAN WANG • *State Key Laboratory of Soil and Sustainable Agriculture, Institute of Soil Science, Chinese Academy of Sciences, Nanjing, China; University of Chinese Academy of Sciences, Beijing, China*
YUXIN WANG • *Division of Natural and Applied Sciences, Duke Kunshan University, Jiangsu, China*
ZIQIN WANG • *Division of Natural and Applied Sciences, Duke Kunshan University, Jiangsu, China*
ARTHUR R. WOLL • *Cornell High Energy Synchrotron Source (CHESS), Cornell University, Ithaca, NY, USA*
ZIJUN XIA • *Division of Natural and Applied Sciences, Duke Kunshan University, Jiangsu, China*

ZHUJUN YAO • *Division of Natural and Applied Sciences, Duke Kunshan University, Jiangsu, China*
YANG YU • *Division of Natural and Applied Sciences, Duke Kunshan University, Jiangsu, China*
HELGA ZELENYÁNSZKI • *Department of Plant Physiology and Molecular Plant Biology, Institute of Biology, ELTE Eötvös Loránd University, Budapest, Hungary*
KELLY ZHENG • *Department of Biology, Amherst College, Amherst, MA, USA*
LU ZHENG • *State Key Laboratory of Soil and Sustainable Agriculture, Institute of Soil Science, Chinese Academy of Sciences, Nanjing, China*
PEIJUN ZHOU • *State Key Laboratory of Soil and Sustainable Agriculture, Institute of Soil Science, Chinese Academy of Sciences, Nanjing, China; University of Chinese Academy of Sciences, Beijing, China*

Chapter 1

High-Throughput Plant Gene Expression Analysis by 384-Format Reverse Transcription-Quantitative PCR for Investigating Plant Iron Homeostasis

Mary Njeri Ngigi and Petra Bauer

Abstract

Analysis of plant gene expression is important in determining iron (Fe) homeostasis gene functions during plant development or in response to biotic and abiotic factors. Reverse transcription-quantitative polymerase chain reaction (RT-qPCR) has many advantages. It is fast, inexpensive, accurate, and reproducible in any lab. Furthermore, RT-qPCR can be scaled up to study several genes of interest in many biological samples from any organism. We hereby provide a straightforward protocol on RT-qPCR analysis using a 384-well format for large-scale gene expression studies on Fe-regulated responses. The protocol highlights in detail, the steps ranging from choice and design of qPCR analysis, collection of plant material and RNA preparation, cDNA synthesis, set up of qPCR and run, thorough analysis of qPCR run data, and display of multiple gene expression data for convenient interpretation.

Key words Iron, Plants, Gene expression, Molecular markers, RNA, cDNA, RT-qPCR, Cq value, Mass standard

1 Introduction

Environmental changes not only impact plant growth but also plant physiology. Plant molecular mechanisms involve genes and proteins that regulate plant responses. Gene expression analysis is a method for the identification of genes involved in plant metabolic pathways. Furthermore, the quantification of transcriptional changes is useful in determining the role of these genes in various biological contexts. Several techniques have been utilized to provide large-scale data on transcriptome profiles during plant interaction with biotic and abiotic factors. For this, whole genome mRNA levels, that are present in a cell at a given time point, are being quantified in a fast and efficient way.

The iron (Fe) nutrition status of a plant can be diagnosed with the aid of well-selected "molecular markers." "Robust" molecular

Jeeyon Jeong (ed.), *Plant Iron Homeostasis: Methods and Protocols*, Methods in Molecular Biology, vol. 2665,
https://doi.org/10.1007/978-1-0716-3183-6_1,

marker genes discriminate the Fe nutritional status across multiple growth conditions and growth stages in *Arabidopsis thaliana* [1]. Analysis of marker genes can be very insightful if they mark co-expressed genes or "co-expression clusters" with well-defined roles in a physiological process. Marker genes of differently regulated co-expression clusters in combination with mutants can be used in identifying hierarchical regulatory connections. Such robust marker genes and differently regulated co-expression clusters are exploited in Fe response studies [1], recently updated in [2], listed in Table 1.

The "High Fe co-expression cluster" steers Fe storage and prevention of Fe-mediated oxidative stress, hence, is a marker for Fe sufficiency in shoots and roots [1]. This sufficient/high Fe-related co-expression cluster contains ferritin genes *FER1* and *FER4*, genes encoding nicotianamine synthase-like *NAS3*, and yellow stripe-like transporter for nicotianamine-Fe complexes such as *YSL1*. Ascorbate peroxidase gene *APX1*, being part of the antioxidant system, is also a suitable marker. These genes are downregulated upon Fe deficiency.

In contrast, two major Fe deficiency-induced co-expression clusters allow resolving Fe deficiency response regulation. One cluster, which we refer to as the *"RED"-Fe co-expression cluster*, comprises genes for internal mobilization of Fe within the plant (e.g., from root to shoot), and regulation in response to low Fe in seedling shoot and root parts [2]. These genes comprise *BHLH38/39/100/101*, *PYE*, all encoding bHLH transcription factors, *OPT3* coding for an Fe signal oligopeptide transporter, *NAS4*, *NRAMP3* encoding an Fe transporter for export of vacuolar Fe, *FRO3* encoding a ferric reductase, and genes encoding small regulatory proteins of the IRON MAN family *IMA1*, *IMA3*, and Fe-dependent E3 ligase of the BRUTUS-type *BTS* [2]. These "RED -Fe markers" are targets of upstream bHLH subgroup IVb and IVc transcription factors, and therefore can be used to assess the activity status of this first low Fe-responsive level of TFs [2]. A second cluster, referred to as the *"BLUE" -Fe co-expression cluster*, is related to seedling root functions for Fe mobilization and acquisition from the rhizosphere [2]. Robust genes of the BLUE cluster are *FRO2* and *IRT1* encoding root Fe reductase and Fe transporter, as well as *F6'H1*, *S8H*, and *CYP82C4* coding for enzymes of the coumarin synthesis pathway [2]. Additionally, *NAS1* and *NRAMP1* are also part of this cluster [2]. The BLUE cluster is induced downstream of the bHLH transcription factor FIT in seedling roots and is useful to assess the activity status of bHLH subgroup Ib and FIT protein [2].

A universally applicable type of gene expression analysis by reverse transcription-quantitative polymerase chain reaction (RT-qPCR) utilizes cDNA generated with polydT primers and fluorescence-based detection of PCR products after each PCR reaction cycle (e.g., via SYBR Green dye). SYBR Green intercalates

Table 1
List of Fe marker genes used for gene expression analysis of Fe responses in Arabidopsis thaliana

Gene name	AGI	STD primers	RT primers
High Fe co-expression cluster			
FER1	AT5G01600	5′ – GCGGCTCAACACTATCCTCT 3′ – ACAGAGCCAACTCCATTGC	5′ – TAAGCCACTACTCCCTCACG 3′ – TTGTTTGTGTCCACCGTAGC
FER4	AT2G40300	5′ – TGGCGTGAAGAAGGATGTGT 3′ – GCGTTTAAACCGGAGCAAACT	5′ – TCTCGTCCCTACCAGCTCTC 3′ – CAAGACCTTTGAGCGCGATG
NAS3	AT1G09240	5′ – GCATGTTCTTCCACACCGTT 3′ – CGTGTTCTGTTTCAGCCCAA	5′ – CAATTGGGAATGTTGGTGG 3′ – TGTTCCTCCCTAGCTCCG
YSL1	AT4G24120	5′ – TATTGTAGCTTACATTTTCGCG 3′ – TATTGTAGCTTACATTTTCGCG	5′ – TTCTTAGCTTCATAGGATCAGTCAA 3′ – GTTCTTGTTCAGAAGTCTACCTGTT
APX1	AT1G07890	5′ – CTACCCAACCGTGAGCGAAG 3′ – TCCTTCTCTCCGCTCAAGAGTTC	5′ – AACCACCTCCAGAGGGTCGT 3′ – GGGCACCAGATAAAGCGACA
"RED" -Fe co-expression cluster			
bHLH38	AT3G56970	5′ – GGAGATAACCTAAATAACGGCA 3′ – GGTCCAGATCAGTGTTAGATTCA	5′ – AGCAGCAACCAAAGGCG 3′ – CCACTTGAAGATGCAAAGTGTAG
bHLH39	AT3G56980	5′ – AACCAAAGCAGCTTCCAAG 3′ – CGAAGAGAAAAAGGACGACA	5′ – GACGGTTTCTCGAAGCTTG 3′ – GGTGGCTGCTTAACGTAACAT
bHLH100	AT2G41240	5′ – GCCTTGCGGAGATCATAGCT 3′ – TCGCGTTGTTCTCCTTCCAA	5′ – AAGTCAGAGGAAGGGGTTACA 3′ – GATGCATAGAGTAAAAGAGTCGCT
bHLH101	AT5G04150	5′ – CATCCCATCAAAGTCTCTCTAGC 3′ – CCTCCAGTCTCACTTTGCAAT	5′ – CAGCTGAGAAACAAAGCAATG 3′ – CAGTCTCACTTTGCAATCTCC
PYE	AT3G47640	5′ – AACGCACCACCTTCTTCTGT 3′ – GTAGCCGAGAAGACCACGAG	5′ – GTTCCCAGGACTTCCCATTT 3′ – GTGTCTGGGGATCAGGTTGT
OPT3	AT4G16370	5′ – TCGGTTATATCCTGCCTG 3′ – GACAGATGTCTCAATAGCTC	5′ – CCCAAACAAGAAGTGGATCCC 3′ – GTGACCAACCAGCTGGCAAT

(continued)

Table 1
(continued)

Gene name	AGI	STD primers	RT primers
NAS4	AT1G56430	5′ – CACTCTCTTCAAGCAGCTCGT 3′ – CTGTAGCAAAAACAGCCAACA	5′ – TGTAATCTCAAGGAAGCTAGGTG 3′ – GCGAACTCCTCGATAATGC
NRAMP3	AT2G23150	5′ – CCTCTTTGGGCTGGTGTTGTT 3′ – TGAGGTAGAGGATGAATGCACCA	5′ – TCGATGTCTTAAACGAGTGGCTT 3′ – CGCGATTGTCTTGTACAAAGGA
FRO3	AT1G23020	5′ – AATCAGATCGACCACCTTGC 3′ – TTCTTTTGGTGAGAAGATTTTGG	5′ – ATCGACCACCTTGCTGTTTC 3′ – TTATCCCACTGCCTCCACTC
IMA1	AT1G47400	5′ – GGCCATCAAGAGATTTGACC 3′ – TGGAAACCATGTTTGTTCATCT	5′ – TGCTTCCACCGTGTATGTTG 3′ – CAGGAGCATAATCATAGCCACTG
IMA3	AT2G30766	5′ – ACGCAGAAGGCAGGCTATAC 3′ – ACAGGCACGATCTACAACTTCA	5′ – GGCAGGCTATACGAATCAACTC 3′ – CGTCACAGTCATCGTCGTCA
BTS	AT3G18290	5′ – AACTTGGATGTTCCCCGTCT 3′ – ATCAACGGGCTTCTTCACAT	5′ – CGGGGAAGGACTAGGAATCG 3′ – CAGCAGATGGGGCAATTTGT
"BLUE" -Fe co-expression cluster"			
IRT1	AT4G19690	5′ – TAGCCATTGACTCCATGGC 3′ – AGAAAACTATGAATCGTGGGG	5′ – AAGCTTTGATCACGGTTGG 3′ – TTAGGTCCCATGAACTCCG
FRO2	AT1G01580	5′ – CCATGCTCGATCTTGTCTTG 3′ – ATTCCGGAACTTTTGAAAGG	5′ – CTTGGTCATCTCCGTGAGC 3′ – AAGATGTTGGAGATGGACGG
IDI1	AT3G07720	5′ – TCGACCTCGAAACTCAGACA 3′ – CGCCATAAACCAACAGACCT	5′ – TCGACCTCGAAACTCAGACA 3′ – AACTCGTTGAGCTCCTGGTG
F6'H1	AT3G13610	5′ – TGGCTCCAACACTCTTGACA 3′ – AAATCGGTTCCTCTCCGTTT	5′ – AAGGCTGCGACTCACAAGTT 3′ – GCTTGCTCTGCAAGAGGACT
S8H	AT3G12900	5′ – ATCGTCGGCTCATGAGTTTT 3′ – ACGTGCGAAGTCGAGAGATT	5′ – ATCGTCGGCTCATGAGTTTT 3′ – TTCCACTCAATCGCCTTCTC

CYP82C4	AT4G31940	5′ – AGAATGGCGGTACAAAACCA 3′ – ATCCTCCGACGATACTGAGC	5′ – AGAATGGCGGTACAAAACCA 3′ – ATCCTCCGACGATACTGAGC
NAS1	AT5G04950	5′ – GTGAAGCAAATCATCGACTTG 3′ – ACTAAACTCCTCGATCATATTC	5′ – TCTCAAAGGCTTTCAACTCTTG 3′ – TACCATTAACCCCGGGC
NRAMP1	AT1G80830	5′ – GGACGGTCTCAATTCATTTC 3′ – GACCAGAAGCAAGAAGCG	5′ – TAATCTTGGTGTTGTCACAG 3′ - TAGGACTTCTCCTGGGTC
Reference gene			
EF1B-α2	AT5G19510	5′ – GCTGCTAAGAAGGACACCAAG 3′ – TGTTCTGTCCCTACTGGATCC	5′ – TATGGGATCAAGAAACTCACAAT 3′ – CTGGATGTACTCGTTGTTAGGC
Genomic DNA contamination control			
EF1B-α2	AT5G19510	5′ – GCTGCTAAGAAGGACACCAAG 3′ – TGTTCTGTCCCTACTGGATCC	5′ – TCCGAACAATACCAGAACTAC 3′ – CCGGGACATATGGAGGTAAG

The genes have been categorized into three clusters. One cluster includes genes that are induced and co-expressed under Fe sufficiency in seedlings and whole plant. The "RED" gene cluster constitutes of genes involved in cellular Fe allocation in response to low Fe and induced in roots and shoots. The "BLUE" gene cluster comprises genes that regulate seedling root Fe acquisition from surrounding soil. Mass standard (STD) and real time (RT) Primer sequences anneal to the indicated genes. Gene clusters were summarized from Ref. [2]

into double-stranded DNA, leading to an exponential increase of fluorescence corresponding to the typical doubling of PCR fragments during PCR. The quantification cycle (Cq) value refers to the cycle number upon which the fluorescence intensity of amplified DNA fragments is significantly higher than the background fluorescence. The Cq value is used for quantification. Cq values can be related to a well-defined initial starting quantity (SQ) of a template provided that a mass standard dilution series is run alongside. Thorough qPCR data analysis is required prior to exporting the data for quantitative gene expression analysis. RT-qPCR has many advantages. It is relatively inexpensive, fast, and easy as it requires only a qPCR device with fluorescence detection. It can be performed by any scientist at any time in the lab. RT-qPCR can be adapted to study any genes of interest (GOIs) in model or non-model organisms, and the choice of genes is flexible. RT-qPCR is a sensitive method that is accurate, reliable, and reproducible across various experimental conditions, such as mutants and treatments. Finally, the method allows for many variations, including automation, multiplexing, and different scales.

We hereby provide a detailed and easy-to-follow protocol for RT-qPCR at large-scale using an SYBR Green-based qPCR detection method in the 384-format. First, deep-frozen harvested plant tissue is processed to yield high-quality total RNA. Then, cDNA is generated by reverse transcription with universal polydT primers and diluted. qPCR is set up using SYBR Green dye and primers for genes of interest (GOIs) and reference control genes together with cDNA and mass standard dilution series. The validity of qPCR data is checked prior to exporting mean Cq and mean SQ values to Excel. Then, SQ values of GOIs are normalized to those of reference genes. Normalization accounts for sample-to-sample variations of RNA and cDNA preparations. The reference gene is important and should be ubiquitously expressed in all analyzed plant tissues regardless of growth conditions, mutants, and treatments. Finally, processed gene expression data are statistically analyzed and displayed in suitable diagrams. In the sections below, we highlight the steps and considerations to be followed from experimental design, and collection of plant material up to the setup and procedure of RT-qPCR analysis. We provide an updated procedure based on [3] using the 384-well format to study large-scale gene expression changes regulated by Fe supply in *Arabidopsis thaliana* seedlings [2].

2 Materials

2.1 General Materials Required for Setting Up a Real-Time RT-qPCR Experiment (See Note 1)

- RNase- and DNase-free PCR-grade water.
- Filter tips for all steps of RNA preparation and elution, cDNA synthesis, and qPCR setup.
- RNase- and DNase-free tubes.

2.2 PCR Reaction Setup for Mass Standard Preparation and qPCR Primer Verification

- PCR reaction mix for regular PCR amplification.
- Oligonucleotide primers.
- 8-strip PCR tubes
- DNA gel electrophoresis reagents (*see* **Note 2**):
  - 50× TAE stock buffer solution:57.1 mL glacial acetic acid in 242 g Tris base dissolved in water, 100 mL of 500 mM EDTA, adjusted to pH 8.0, and 1 L of final volume
  - Agarose powder for 1% agarose gel.
  - DNA staining dye.
- DNA molecular weight marker for size determination (Gene Ruler™ 1 kb DNA Ladder or equivalent).
- DNA gel extraction kit.
- UV spectrometer (*see* **Note 3**).

2.3 Sample Preparation and RT-qPCR

- RNA isolation: RNeasy plant mini kit (Qiagen) or equivalent.
- DNase I buffer.
- DNase I.
- Ribolock RNase Inhibitor.
- EDTA.
- Oligo $(dT)_{18}$ primer.
- cDNA synthesis: RevertAid First cDNA synthesis kit,
- RNase- and DNase-free PCR-grade water.
- Quantitative real-time PCR: iTaq™ Universal SYBR ® Green Supermix (Bio-Rad) or equivalent, aliquoted in 1.5 mL tubes from stock and stored at −20° in darkness.
- Precision Blue™ Real-Time PCR Dye DNA oligos for PCR amplification.
- Colored 8-strip PCR tubes for cDNA storage.
- 384-well qPCR microplate
- Adhesive optical sealing tape adapted for 384-well microplates.
- Real-time Thermal Cycler C1000 Touch CFX384 or equivalent.

3 Methods

The generation of reproducible qPCR results is guided by a careful choice of several parameters. The following section highlights considerations during the planning and execution of qPCR experiments for gene expression: (1) design and optimization of qPCR oligonucleotide primers for high PCR efficiency, (2) mass standard preparation, (3) evaluation of RT qPCR primers, (4) steps and considerations for the generation of plant material and harvesting, (5) RNA preparation and cDNA synthesis, (6) qPCR setup, (7) qualitative analysis of qPCR run and gene expression data visualization. A summary of this experimental workflow is illustrated in Fig. 1.

3.1 Design and Optimization of qPCR Oligonucleotide Primers

1. Select genes of interest and reference gene. Obtain genomic nucleotide sequences from https://seqviewer.arabidopsis.org/. Obtain coding sequence and cDNA sequence from https://www.arabidopsis.org/.
2. Analyze cDNA sequence information using DNA analysis software for the following parameters: conserved regions in cDNA sequences among gene family members, 5′ and 3′ untranslated regions (UTRs), start and stop codons of translated regions as well as splicing and alternative splicing sites.
3. Test different reference gene candidates for different physiological treatments and plant species (*see* **Note 4**).
4. Design 5′ and 3′ qPCR primers (= RT primers) for the target sequence using Primer3 https://primer3.ut.ee and check primer-binding specificity by performing a BLAST search of primer sequences against an available sequence database of transcribed gene sequences of species or close relatives and ensure that no off-target sites are present (*see* **Notes 5** and **6**).
5. Design 5′ and 3′ mass standard primers (= STD primers) to generate a linear mass standard product of approximately 1 kb that encloses the exact qPCR target region (*see* **Note 7**).

3.2 Mass Standard Preparation

1. Mass standards can be amplified using cDNA or gDNA as a template. Perform PCR with 1 μL template and 1 μL of 10 μM 5′ and 3′ STD primers (*see* **Note 8**).
2. Run 1% Agarose gel electrophoresis using 15 μL of PCR reaction to verify the product size of the mass standard DNA fragment. Using a razor blade and under visible UV light cut out the agarose gel area containing the band.
3. Purify the DNA fragment from the gel using a commercial gel extraction kit and measure the concentration of the final eluate.

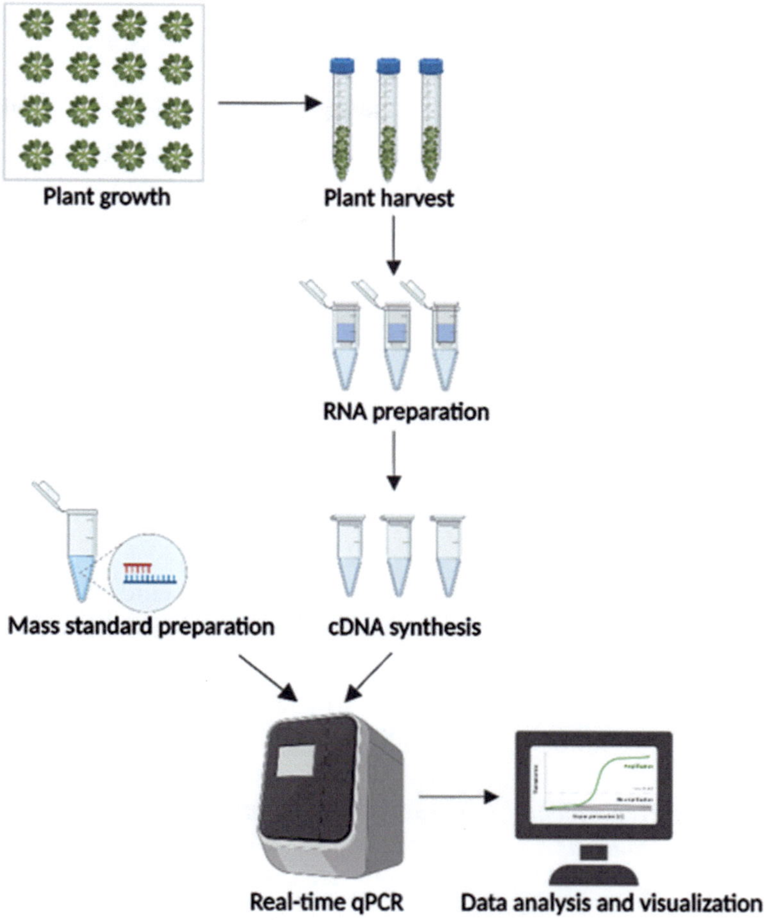

Fig. 1 Experimental workflow for gene expression analysis by RT-qPCR. Workflow from plant growth to qPCR data analysis and visualization. Plants are grown under the study conditions in three biological replicates. Different plant materials are harvested followed by RNA preparation, cDNA synthesis and real-time qPCR to determine absolute expression levels via mass standard curve analysis. Data are qualitatively and quantitatively analyzed and appropriate diagrams used for visualization of results. Figure prepared with Biorender.com

4. Determine and check the integrity of the DNA sequence of the purified fragment by Sanger sequencing, using the service and recommendation of, e.g., a commercial supplier.
5. Determine the molecular concentration of purified mass standard DNA using a molecular conversion tool http://molbiol.edu.ru/eng/scripts/01_07.html and calculate DNA copy number/μL solution. Adjust final molar concentration to 10^9 DNA copies/10 μL. Store at −20 °C.
6. Prepare a serial dilution of DNA mass standards in 1.5 mL reaction tubes by subsequent 1:10 dilution steps to obtain 10^8, 10^7, 10^6, 10^5, 10^4, 10^3, 10^2/10 μL solution. Prepare

800–1000 μL of each mass standard dilution and aliquot ca. 50 μL into pre-labeled 8-strip PCR tubes from 10^7 to 10^2 (*see* **Note 9**).

3.3 Evaluation of RT-qPCR Primers

1. Amplify 1 μL of diluted (10^7 copies/10 μL) mass standard DNA and of diluted cDNA with 5′ and 3′ RT-qPCR primers in a regular 40-cycle PCR reaction.
2. Run 2% agarose gel electrophoresis using 15 μL PCR product to verify the presence and size of the expected DNA fragments (*see* **Note 10**).
3. RT primers are usually used in a 1:1 ratio, but varying from 3:1, 2:1, 1:1, 1:2, and 1:3 (primer matrix) allows optimization (*see* **Note 11**).
4. The qPCR fragments must be sequenced to verify their expected identities. Design new primers if not successful.

3.4 Plant Material Generation and Harvesting

1. Grow plants in three independent biological replicates and under the study conditions (Fig. 1) (*see* **Note 12**).
2. Harvest plant materials (*see* **Note 13**), immediately shock-freeze with liquid nitrogen, and if needed store at −80 °C until use.

3.5 RNA Preparation and cDNA Synthesis

1. Grind frozen plant material to a fine powder with liquid nitrogen, or use an electrical homogenizer stick or a homogenization device such as Precellys (*see* **Note 14**).
2. Using a maximum of 100 mg of plant material, isolate RNA following the manufacturer's instructions (*see* **Notes 15** and **16**).
3. Elute RNA in 50 μL RNase-free water. In case of insufficient material, 30 μl RNase-free water is sufficient. All steps during RNA isolation should be carried out under RNase-free conditions to avoid RNA degradation by RNases (*see* **Note 17**).
4. Measure RNA concentration at A260/280.
5. In 200 μL PCR tubes, add 500–1000 ng RNA and mix with ultrapure RNase- and DNase-free water to obtain 7.75 μL total volume in all samples. Additionally, include a negative cDNA control with water but no RNA included (*see* **Note 18**).
6. Prepare "DNase I master mix" containing per reaction: 1 μL of DNase I buffer, 1 μL of DNase I (1 U/μL), and 0.25 μL of Ribolock RNase inhibitor (40 U/μL). Add 2.25 μL of the master mix to 7.75 μL RNA to make a total volume of 10 μL. Incubate reaction mix at 37 °C for 30 min in a PCR thermocycler.

7. To deactivate the DNase, prepare "DNase deactivation and oligo dT primer annealing" mix containing per reaction 1 μL of 50 mM EDTA and 1 μL of 0.5 μg/μL oligo $(dT)_{18}$ primer. Add 2 μL of the DNase deactivation and oligodT primer annealing master mix to the reaction volume from **step 6** for a total reaction mix of 12 μL. Incubate at 65 °C for 10 min, then at 4 °C until the next step.
8. To synthesize cDNA, prepare "reverse transcription" master mix (*see* **Note 19**): mix 0.5 μL of RNase- and DNase -free water, 4 μL of reverse transcriptase buffer, 0.5 μL of Ribolock RNase inhibitor (40 U/μl), 2 μL of 10 mM (each) dNTP mix and 1 μL of reverse transcriptase (200 U/μl). Add 8 μL of the cDNA synthesis mix to the reaction volume from **step 7** for a total reaction volume of 20 μL. Incubate at 42 °C for 1 h, then at 72 °C for 10 min.
9. Combine 20 μL of each sample reaction from **step 8** with 180 μL RNase- and DNase -free PCR-grade water and mix well. Store these diluted cDNA stocks at −20 °C until further use (*see* **Note 20**).
10. Prepare another 1:10 dilution of diluted cDNA samples for real-time qPCR.
11. In a 1.5 μL reaction tube combine 270 μL of RNase- and DNase-free PCR-grade water and 30 μL cDNA from **step 9**, then mix well.
12. Aliquot 30 μL of these double-diluted cDNA samples into 8-strip PCR tubes and store at −20 °C until further application.

3.6 qPCR Setup

Proceed to qPCR reaction setup for the mass standards, the double diluted cDNA samples, the negative cDNA control (water), and qPCR negative control (NC) in technical replicates.

1. Proceed to qPCR reaction setup for the mass standards, the double diluted cDNA samples, the negative cDNA control (water), and qPCR negative control (NC) in technical replicates.
2. Design and program the qPCR parameters and plate setup using the CFX Maestro™ Software (Bio-Rad) or any available software compatible with 384-well qPCR microplates (Figs. 2 and 3; *see* **Note 21**). Two technical replicates have to be performed for each sample.
3. Thaw the frozen mass standards, cDNA, and negative control samples in the PCR strips and briefly spin down.
4. Prepare a qPCR master mix by adding the following volumes per reaction (*see* **Note 22**): 4.8 μL of Universal SYBR ® Green

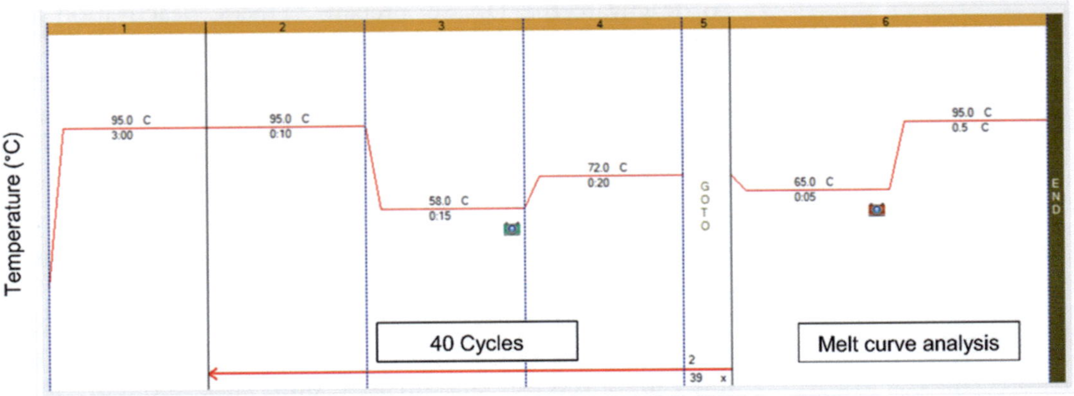

Fig. 2 qPCR program. The program consists of an initial long denaturation, followed by 40 cycles of PCR and melt curve data collection. The camera indicates fluorescence recording. (Figure was generated by CFX Maestro™ Software version 2.3 (Bio-Rad))

Supermix, 0.1 μL of each 10 μM 5′ and 3′ oligonucleotide primer, 0.025 μL of Precision Blue real-time PCR dye (Bio-Rad) or equivalent indicator for precision pipetting (Fig. 3). Mix carefully by vortexing.

5. Using a multi-dispenser pipette add 5 μL of the qPCR master mix into each well bottom of the 384-well qPCR microplate (Fig. 3). Confirm that every well is loaded with the master mix (*see* **Note 23**).
6. Using a 10 μL multichannel pipette transfer 5 μL of the cDNA samples, mass standards, water, and negative controls from the 8-strip PCR tubes into the loaded wells with master mix (Fig. 3; *see* **Note 24**).
7. Cover the microplate using optical sealing tape. Gently tap the plate against the bench top then briefly spin down using a plate centrifuge. The loaded 384-well qPCR plates can be stored for a few hours in the darkness under 4 °C (*see* **Note 25**).
8. Load the qPCR microplate into the qPCR machine and start the program. Ensure to save the data file after a run is complete.

3.7 Qualitative Analysis of qPCR Run and Gene Expression Data Visualization

1. Verify a mass standard curve and PCR efficiency (Fig. 4). High initial template quantities have low Cq values while low abundance templates are detected at later time points hence high Cq (>25) (Fig. 4a; *see* **Note 26**).
2. Check the melt curves of all cDNA sample runs and compare with those of the mass standard samples (Fig. 5; *see* **Note 27**).
3. Verify technical duplicate amplification data for each sample, a recommended difference is <0.5 Cq value.
4. Analyze the data following the steps in (Fig. 6):

RT-qPCR Master Mix preparation	1 reaction	336 reactions
Universal SYBR ® Green Supermix	4.8 μL	1612.8 μL
30 μM Forward primer (5`)	0.1 μL	33.6 μL
30 μM Reverse primer (3')	0.1 μL	33.6 μL
Precision Blue real-time PCR dye	0.025 μL	8.4 μL
Total volume	**5.025 μL**	**1688.4 μL**

I.

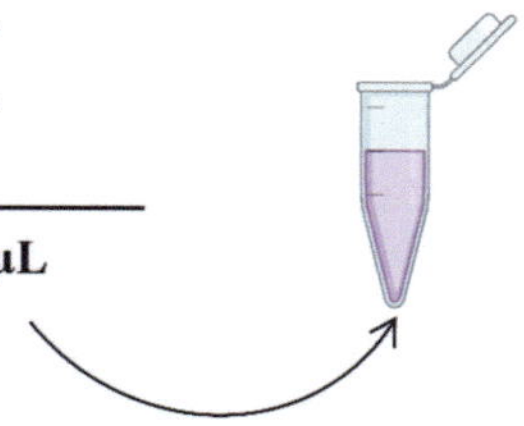

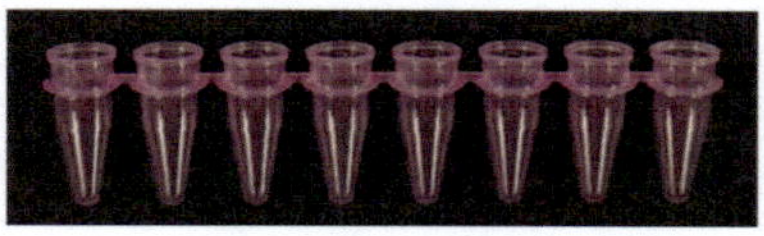

II.

5 μL double-diluted cDNA per well

	1	2	3	4	5	6	7	8	9	10	11	12	13	14	15	16	17	18	19	20	21	22	23	24
A	C1	C1	C1	C1	C1	C1	C1	C1	C1	C1	C1	C1	C1	C1	C1	C1	C1	C1	C1	C1	C1	C1	C1	C1
B	C2	C2	C2	C2	C2	C2	C2	C2	C2	C2	C2	C2	C2	C2	C2	C2	C2	C2	C2	C2	C2	C2	C2	C2
C	C3	C3	C3	C3	C3	C3	C3	C3	C3	C3	C3	C3	C3	C3	C3	C3	C3	C3	C3	C3	C3	C3	C3	C3
D	T1	T1	T1	T1	T1	T1	T1	T1	T1	T1	T1	T1	T1	T1	T1	T1	T1	T1	T1	T1	T1	T1	T1	T1
E	T2	T2	T2	T2	T2	T2	T2	T2	T2	T2	T2	T2	T2	T2	T2	T2	T2	T2	T2	T2	T2	T2	T2	T2
F	T3	T3	T3	T3	T3	T3	T3	T3	T3	T3	T3	T3	T3	T3	T3	T3	T3	T3	T3	T3	T3	T3	T3	T3
G	O	O	O	O	O	O	O	O	O	O	O	O	O	O	O	O	O	O	O	O	O	O	O	O
H	10^7	10^7	10^7	10^7	10^7	10^7	10^7	10^7	10^7	10^7	10^7	10^7	10^7	10^7	10^7	10^7	10^7	10^7	10^7	10^7	10^7	10^7	10^7	10^7
I	10^6	10^6	10^6	10^6	10^6	10^6	10^6	10^6	10^6	10^6	10^6	10^6	10^6	10^6	10^6	10^6	10^6	10^6	10^6	10^6	10^6	10^6	10^6	10^6
J	10^5	10^5	10^5	10^5	10^5	10^5	10^5	10^5	10^5	10^5	10^5	10^5	10^5	10^5	10^5	10^5	10^5	10^5	10^5	10^5	10^5	10^5	10^5	10^5
K	10^4	10^4	10^4	10^4	10^4	10^4	10^4	10^4	10^4	10^4	10^4	10^4	10^4	10^4	10^4	10^4	10^4	10^4	10^4	10^4	10^4	10^4	10^4	10^4
L	10^3	10^3	10^3	10^3	10^3	10^3	10^3	10^3	10^3	10^3	10^3	10^3	10^3	10^3	10^3	10^3	10^3	10^3	10^3	10^3	10^3	10^3	10^3	10^3
M	10^2	10^2	10^2	10^2	10^2	10^2	10^2	10^2	10^2	10^2	10^2	10^2	10^2	10^2	10^2	10^2	10^2	10^2	10^2	10^2	10^2	10^2	10^2	10^2
N	NC	NC	NC	NC	NC	NC	NC	NC	NC	NC	NC	NC	NC	NC	NC	NC	NC	NC	NC	NC	NC	NC	NC	NC

GOI1 GOI2 GOI3 GOI4 GOI5 GOI6 GOI7 GOI8 GOI9 GOI11 GOI12 REF

5 μL mass standards per well

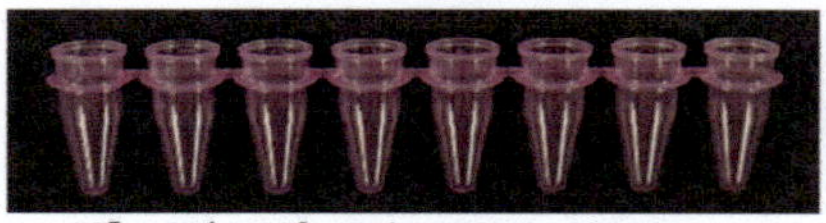

III.

10^7 10^6 10^5 10^4 10^3 10^2 NC

Fig. 3 qPCR setup. The order of qPCR pipetting is shown as the circled numbers I, II and III. Top, preparation of the qPCR master mix with an example of calculations using a 384-well microplate system. Master mixes corresponding to each gene of interest (GOI) are pipetted in the GOI positions (1–22) and reference gene on position 23 and 24 in the plate. 5 μL of the master mix is transferred from a 1.5 mL reaction tube into the respective wells. The plate pipetting scheme shows positions of unknown cDNA samples. All cDNA samples are pipetted in two technical duplicates, 3 biological replicates for control group (C1–C3) and treatment group (T1–T3). 5 μL of cDNA templates are transferred from the 8-strip PCR tubes into the microplate wells using a multichannel pipette. Well positions with "0" were loaded with the negative water cDNA control. 5 μL of serially diluted mass standards (107–102) corresponding to the GOIs and reference genes are added at shown well positions. A negative qPCR control (NC) without template is included. Figure prepared using Biorender.com

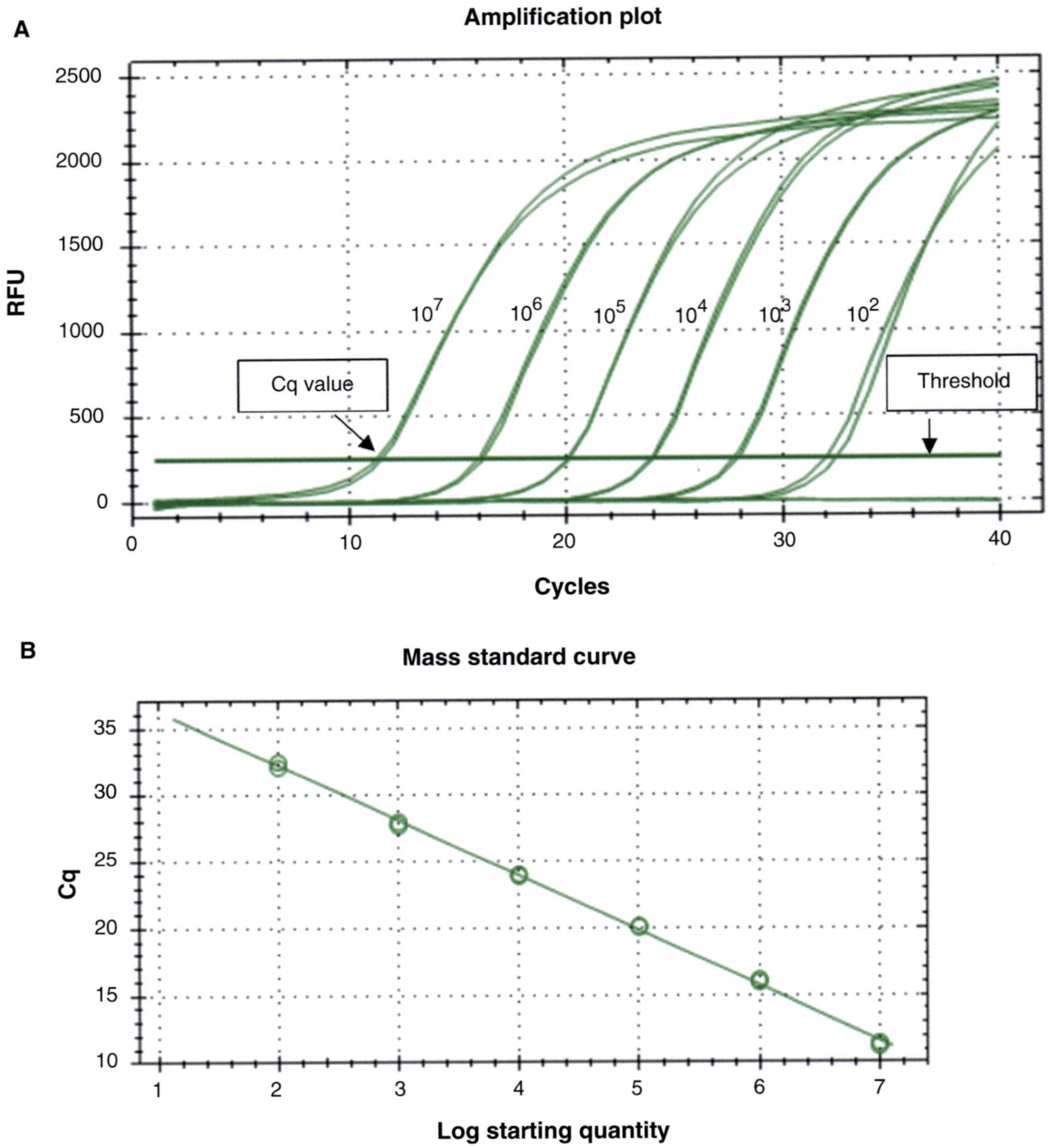

Fig. 4 Mass standard curve analysis method for qPCR-based determination of initial starting quantities of templates (SQs) by absolute quantification. (**a**) Amplification plots of serially diluted mass standards in the range 107 to 102 DNA copies per reaction. Each mass standard qPCR reaction was run as a technical duplicate. The fluorescence is recorded during the qPCR run cycles as Relative Fluorescence Units (RFU) and measured at the end of each cycle. Quantification cycle (Cq) value indicates a time point at which the RFU of qPCR products reaches a set threshold level. (**b**) Mass standard curve derived from (**a**) showing a linear inverse correlation between template initial starting quantities and Cq measurements. The qPCR efficiency is calculated from this curve. (Figures were generated by CFX Maestro™ Software version 2.3 (Bio-Rad))

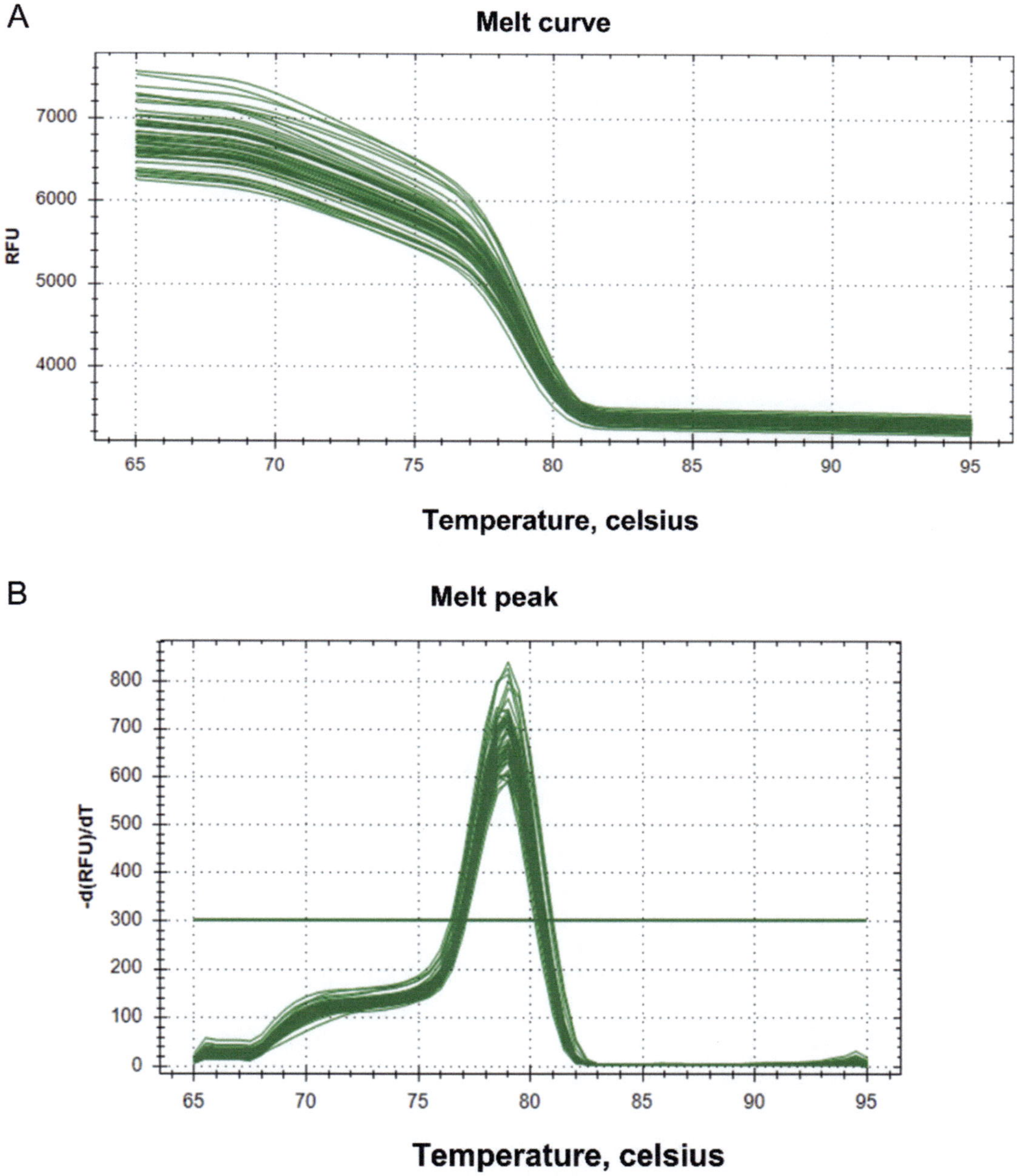

Fig. 5 Quality control of qPCR run data. (**a**) Melt curve analysis of mass standard and cDNA samples of a GOI example amplification. The number of relative fluorescence units (RFU) decreases with increasing temperature due to melting of PCR products leading to a drastic loss of fluorescence. (**b**) Melt peak analysis of (A). A single melt peak suggests melting of qPCR products at a specific temperature indicating that a single PCR product was present in all samples. (Figures were generated by CFX Maestro™ Software version 2.3 (Bio-Rad))

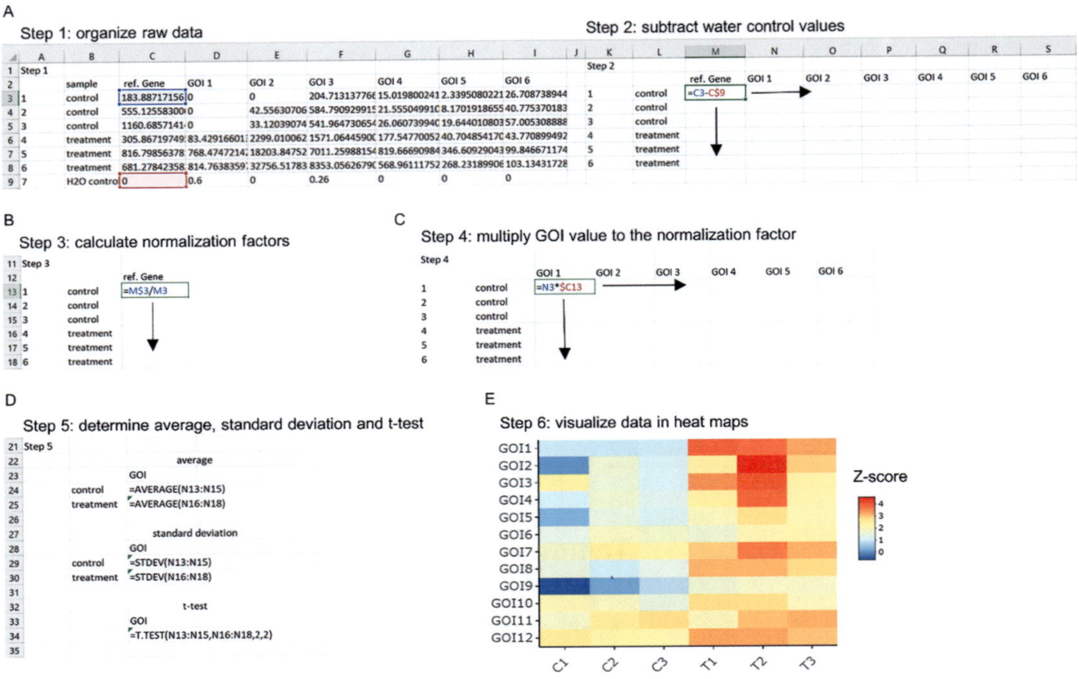

Fig. 6 Data analysis and visualization of normalized gene expression values, corresponding to normalized initial starting quantities (SQ values). (**a**) qPCR data are exported to an excel file. Indicated initial SQ values are an average of the technical duplicates for 6 biological samples (3 control, 3 treatment groups), 6 GOIs and a reference gene (*EF1B-ALPHA2*) including a negative cDNA water control (H_2O control). Water contamination values are subtracted from each gene, as indicated. (**b**) Normalization factors of the reference gene are calculated as indicated. (**c**) The absolute expression values of each gene of interest are normalized with calculated normalization factors, as indicated. (**d**) Final results are subjected to statistical analysis including average and standard deviation of the three biological replicates and performing unpaired two-tailed t-test to determine the statistical significance difference between control and treatment group. More than two biological sample groups require other statistical tests. (**e**) Heat map diagram showing expression changes of 12 GOIs tested under biological replicates of control conditions (C1–C3) and treatment conditions (T1–T3). The heatmap was generated in R studio version 3.8 using calculated Z-scores of normalized absolute mean gene expression values. Red colors indicate higher z scores, hence gene up-regulation, while blue colors indicate lower scores and gene down-regulation

(a) First, export the qPCR data in an excel spreadsheet. Add appropriate sample names and delete rows and columns that are not necessary to retain mean Cq values and mean SQ values, whereby the mean values are those of technical duplicates (Fig. 6a).

(b) Subtract the negative cDNA water control values from every sample (Fig. 6a). Also subtract the genomic DNA values in case of primer amplification in a genomic region. Genomic DNA amplification can be controlled by testing the amplification of an internal genomic DNA control fragment, such as in the genomic non-transcribed region *EF1B-ALPHA2*.

(c) Calculate normalization factors for all samples based on the SQ values of the reference gene such as *EF1B-ALPHA2* (e.g., set the value of one sample to 1 and calculate the factors by which all other SQ sample values need to be multiplied/divided to correspond accordingly to 1) (Fig. 6b).

(d) Proceed to normalization of each GOI SQ value by applying the respective normalization factors of each sample. This yields normalized expression values (Fig. 6c; *see* **Note 28**).

(e) Determine the mean normalized expression values for biological replicates, corresponding standard deviations, and conduct appropriate statistical tests (Fig. 6d).

(f) Appropriately, visualize final data using diagrams or heatmaps (Fig. 6e; *see* **Note 29**).

4 Notes

1. Prepare all reaction solutions using RNase- and DNase-free PCR-grade water. To avoid sources of contamination, use filter tips during all steps of RNA preparation and elution, cDNA synthesis, and setting up qPCR. Set up reactions in RNase- and DNase-free tubes. Unless otherwise stated, store all solutions and kits at room temperature.
2. Dilute 50× TAE stock to 1× TAE solution, and boil 100 mL 1× TAE buffer with 1 g Agarose powder for 1% agarose gel, add 3 μL DNA staining dye. Pour gel and allow to cool down.
3. Measure DNA/RNA concentration using a UV spectrometer (Tecan microplate reader or equivalent). OD_{260} corresponds to DNA and RNA maximum absorption spectra while OD_{280} corresponds to the one of proteins. Therefore, $OD_{260:280}$ ratio provides the degree of RNA purity and should be close to 2.0.
4. Choose a reference gene that is reliable in terms of showing similar expression in different plant tissues regardless of the growth conditions. In case of increased variability of reference gene expression among the collected materials due to the biological nature of the organ sources or other growth factors, use two reference genes and calculate an average of them before normalization of the genes of interest values. Different reference gene candidates must be tested for different physiological treatments and plant species. In some cases, it may be necessary

to average the expression profile of up to three reference genes. We found that the elongation factor gene *EF1B-ALPHA2* (AT5G19510) yields the best results with little biological sample-to-sample variation in *Arabidopsis thaliana* Fe deficiency response regulation studies in seedlings.

5. Carefully analyze gene sequences when designing primers. Identify unique regions within the cDNA and coding sequences which may affect primer target binding. If possible, design primers in the coding sequence when the gene of interest exists as a single copy. Avoid the regions that form secondary structures (e.g., hair pins). In the case of multiple copy genes, primers can be designed in 5′ and 3′ UTRs of the cDNA as these regions are transcribed. Preferentially select primers that bind towards the 3′ end of the cDNA.
6. A convenient tool for primer design with BLAST function is provided at https://www.ncbi.nlm.nih.gov/tools/primer-blast/. Ideally, oligonucleotide primers are 18–23 bp long, 50–60% GC content, 60 °C annealing temperature, not self-complementary, form no hair pins, and amplify between 80–150 bp qPCR products. Avoid shorter than 80 bp products as they are not easily distinguishable from primer dimers and generally reduce the PCR efficiency.
7. Mass standards serve as external calibration for amplified cDNA fragments and can be designed to span the qPCR and surrounding genomic DNA or cDNA region.
8. Plasmid DNA and cDNA can also be used as templates for mass standard amplification.
9. Prepare mass standard stocks for each gene in advance and aliquot small amounts into 1.5 mL tubes for storage at −20 °C. Thaw the required amounts upon use and avoid freeze-thawing of the solutions. The 10^7 to 10^2-diluted DNA fragments are not stable at −20 °C and it is advised to prepare the dilutions shortly before use. During the setup of RT-qPCR, we recommend using various colored 8-strip PCR tubes to identify different mass standards for every gene.
10. Testing RT primers before the real-time qPCR experiment is a crucial step in minimizing off-target primer binding. If there is non-specific amplification or no PCR products try to maximize the parameters of the PCR reaction (e.g., use the lowest primer concentration), adjust Mg^{2+} concentration and PCR annealing temperature.
11. Diluted qPCR primers and primer stock solutions should be stored at −20 °C. Optimize the primer concentrations for increased PCR efficiency. Generally, lower primer concentrations lead to increased specificity and reduced off-target amplification.

12. Minimize differences in growth conditions of biological replicate samples and variations in harvesting time. Some Fe-regulated genes are responsive to the circadian rhythm. Take note of the harvesting time points (e.g., always harvest 3–5 h after the onset of light).
13. Pool roots and shoots of multiple seedlings (e.g., 10 seedlings grown for 14 days at +Fe followed by 3 days on −Fe is sufficient for one biological sample). Pooling from at least 10 different seedlings is useful to reduce sample-to-sample variation.
14. Grind plant material by mortar and pestle or using an automated tissue homogenizer stick that can be inserted in a tube. In case of a high sample number, better use an automated Precellys machine for more reproducible and quick tissue homogenization. It is important to shock freeze plant material using liquid nitrogen before the grinding steps. 2 mL screw cap tubes are safer to work with in liquid nitrogen. Add 450 μL of extraction buffer to the tubes already containing 1.4 mm Precellys ceramic beads before proceeding with tissue homogenization via the Precellys device at 6500 rpm for 20 s and 3 cycles. The number of beads in the tubes can be varied depending on sample amount (approximately 20 beads per 70 mg plant sample).
15. The amount of used plant material affects the yield of RNA. We recommend using 50–70 mg of starting material for RNA extraction. A higher amount of material such as leaf samples may lead to clogging of columns, low yield, impurity, and DNA contamination.
16. For miniature samples such as embryos and root or shoot apices, we advise preserving the RNA quality by using a stabilizing reagent like RNA*later*® [4]. Pellet the tissues and carefully pipette out this solution before adding extraction buffer and tissue homogenization.
17. RNA is highly unstable and easily degraded by ubiquitous RNases. RNA extraction buffers may contain RNase-deactivating compounds like guanidium thiocyanate (GTC), β-mercaptoethanol, and phenolic substances which are toxic. Always wear gloves. Generally, RNA is protected in the extraction buffers and under deeply frozen conditions. It is recommended to continue working under RNase-free conditions during subsequent steps after RNA extraction. Working bench surfaces and materials should be wiped clean, always use filter tips during RNA purification steps and upon elution, and change gloves in case of contamination. Carefully open tubes with two fingers on the tube cap and avoid touching the inner tube surfaces to reduce RNase contamination from

fingers. Store RNA at −20 °C in case of short-term use and −80 °C for long-term storage.

18. Before proceeding to cDNA synthesis, it is important to eliminate genomic DNA contamination by DNase treatment. This helps to ensure that only cDNA amplification takes place during the RT-qPCR. Label 200-μL-PCR strips with respective sample names.

19. Perform cDNA synthesis steps in 200 μL 8-strip PCR tubes. Vigorous sample mixing in the 8-strip tubes (e.g., by vortexing) should be avoided due to the risk of sample spillover and cross-contamination which may hugely bias the gene expression data. It is mandatory to perform a 1:10 dilution of cDNA stocks in 1.5 mL reaction tubes which should be firmly closed before vortexing. During the cDNA dilution series, it is more accurate and easier to pipet large volumes such as 30 μL (hence add 270 μL of water to 30 μL cDNA). The 384-well system requires a 5 μL cDNA sample per well, hence a 50 μL volume of double-diluted cDNA is sufficient for four qPCR runs i.e., 10 μL for two technical duplicates × 4 genes. Prepare up to five 8-strip tubes of the double diluted cDNA for convenience of pipetting during the qPCR setup. We recommend using various colored strips for ease of identifying different cDNA samples (e.g., samples 1–8 in red strips, 9–16 in green strips, and 17–24 in yellow strips).

20. cDNA samples can be stored long-term (up to 6 months) under −20 °C in reaction tubes clearly labeled with "date of cDNA synthesis," "sample name," and "sample number." Keep tubes in sealable plastic bags labeled with printout of the experimental setup, sample description, and date.

21. Prepare a plate setup of the 384-well beforehand to determine the number of samples that can fit in the well given the number of genes tested. It is possible to combine different genes in one 384-well plate when a few samples are to be tested. In this case, consider that genes combined on one plate have similar primer annealing temperatures. For higher sample amounts, prepare only one plate per gene and in case the samples are more than the well number include a commercially available plate-to-plate calibration standard (e.g., from TATAA Biocentre) to account for variability in the different plates.

22. For each 100 qPCR reaction, calculate and prepare 10 μL extra qPCR mix to avoid shortage upon pipetting.

23. Pipette qPCR master mix directly into well bottom of the 384 PCR plate using a 10 μL pipette or take up a higher volume and in several steps dispense with a 5 μL multi-step pipette.

24. Use a multichannel 10 μL pipette to add the cDNA samples without touching the master mix. For this hold the pipette at a specific angle and dispense the solution inside the wells. Reliable and reproducible qPCR data are dependent on accurate pipetting. Therefore, higher-volume pipets (>10 μL) are not suitable for setting up 10 μL qPCR reactions.
25. Cover the microplate using optical sealing tape to avoid evaporation of the reaction solutions in the plate wells. Firmly press the tape onto the plate by carefully pressing from the middle to the edges of the microplate to avoid uneven surfaces on the seal.
26. A mass standard curve is a regression line which shows an inverse correlation between template fluorescence and the Cq value (Fig. 4b). PCR efficiency is extrapolated from the mass standard curve whereby the doubling of PCR products at every cycle is calculated. A PCR efficiency of 95–100% indicates optimized PCR conditions. Samples with multiple peaks or aberrant curves must be excluded from analysis. Mass standard values that are off the linear curve can be excluded from analysis. At best PCR efficiency should be 100% or close to this, hence an indication of product doubling with every PCR cycle. PCR efficiencies below 80% or exceeding 110% need to be optimized for PCR conditions such as primer molarity and the annealing temperature of a given primer combination. Design new primers in case of several and consistently aberrant efficiencies. The technical duplicates should not differ by more than 0.5 Cq values to be considered reliable. However, in the case of high Cq values (>30) due to low SQs, higher than 0.5 Cq variation between technical duplicates is acceptable. Aberrant Cq values have to be excluded from analysis.
27. A melt curve indicates the presence of double-stranded DNA and is obtained by increasing the temperature from 60 to 94 °C in small incremental steps and recording the fluorescence after each step. The fluorescence will drastically drop when the melt temperature of the PCR fragment is reached (Fig. 5a). The melt curve is often represented as a change in fluorescence per incremental unit of temperature increase, yielding a peak that reflects the melt temperature, and a single melt curve peak indicates the presence of a single qPCR product (Fig. 5b). The melt curves must be identical between cDNA and mass standard samples. Samples with multiple peaks or aberrant curves must be excluded from analysis.
28. Upon data processing, only negligible SQ values are acceptable for negative water cDNA controls and genomic DNA contamination.

29. For data visualization of high numbers of biological samples, experimental conditions, and analyzed genes, we recommend using a heat map for display as this provides an overview of all gene expression changes in the samples under control and treatment conditions. In such situations, perform a two-tailed ANOVA statistical test for data statistical analysis.

References

1. Ivanov R, Brumbarova T, Bauer P (2012) Fitting into the harsh reality: regulation of iron-deficiency responses in dicotyledonous plants. Mol Plant 5(1):27–42. https://doi.org/10.1093/mp/ssr065
2. Schwarz B, Bauer P (2020) FIT, a regulatory hub for iron deficiency and stress signaling in roots, and FIT-dependent and -independent gene signatures. J Exp Bot 71(5):1694–1705. https://doi.org/10.1093/jxb/eraa012
3. Ben Abdallah H, Bauer P (2016) Quantitative reverse transcription-qPCR-based gene expression analysis in plants. Methods Mol Biol 1363:16. https://doi.org/10.1007/978-1-4939-3115-6
4. RNAlater ® Tissue Collection: RNA Stabilization Solution. Thermo Fish Sci (2014)

Chapter 2

The Use of Spectral Imaging to Follow the Iron and pH-Dependent Accumulation of Fluorescent Coumarins

Kevin Robe, Geneviève Conjero, and Christian Dubos

Abstract

Plants challenged with iron deficiency produce in their roots and secrete into the rhizosphere several small molecules named coumarins that derive from the phenylpropanoid pathway. Coumarins are biosynthesized in different root cell types and transported to the root epidermis prior to their secretion in the surrounding media. Taking advantage of the natural fluorescence of most coumarins glycosides when exposed to UV light, we developed a method to uncover their individual cellular localization and accumulation. This approach couples spectral imaging acquisition and linear unmixing analysis. In this protocol, we describe guidelines, experimental setup, and conditions for the analysis of coumarins localization and accumulation in *Arabidopsis thaliana* root seedlings grown in control and iron deficiency conditions, at both acidic and alkaline pH.

Key words Spectral imaging, Linear unmixing, Scopolin, Fraxin, Esculin, Iron, pH

1 Introduction

Although abundant in most soils, iron (Fe) acquisition is challenging for several plant species. To meet their nutritional needs in Fe, plants have evolved elaborate strategies [1]. Among them, the secretion of Fe binding and mobilizing coumarins into the rhizosphere by the plant root system has emerged as a critical component of the Fe acquisition strategy of non-grass species, such as *Arabidopsis thaliana*, especially in alkaline soils [2]. Coumarins are secondary metabolites derived from the phenylpropanoid pathway [3–6]. Previous studies highlighted that Arabidopsis roots secrete several coumarins, such as scopoletin, fraxetin, esculetin, and sideretin, via the ABCG37/PDR9 (PLEIOTROPIC DRUG RESISTANCE 9) transporter [7].

HPLC and mass spectrometry analysis have been widely used to quantify coumarin content in Arabidopsis roots. For instance, it was shown that even though Fe deficiency induces the production of

Jeeyon Jeong (ed.), *Plant Iron Homeostasis: Methods and Protocols*, Methods in Molecular Biology, vol. 2665,
https://doi.org/10.1007/978-1-0716-3183-6_2,

coumarins, the pH of the growth substrate is also a strong driver of coumarins biosynthesis [1, 8, 9]. For instance, fraxetin is the main binding and mobilizing coumarin produced under alkaline pH whereas sideretin is favored under acidic conditions [3]. These technics are powerful to study the global response of the plant; however, they do not provide any information on coumarins localization within the root tissues and/or cell types.

It is noteworthy that in roots, coumarins are glycosylated prior to their storage in the vacuole [9]. Interestingly, the main coumarins found in Arabidopsis roots are fluorescent when exposed to UV light once glycosylated, except the sideretin glucoside. These coumarins display overlapping but different emission spectra after exposure to UV light, a feature we exploited to study the *in planta* distribution and accumulation of scopolin, esculin, and fraxin in response to environmental stresses, such as Fe availability and pH level of the growth media. For this purpose, the use of spectral imaging coupled with linear unmixing is a very insightful tool. This microscopy technique enables the simultaneous detection of fluorescent molecules with highly overlapping emission spectra [10–13]. This is because spectral imaging allows the collection of emission spectrum information at each pixel of a given sample and separates signals from multiple fluorescent molecules.

In this chapter, we describe the guidelines, experimental setup, and conditions to study the localization and the accumulation of coumarins in Arabidopsis root seedlings grown in media with contrasted Fe availability and pH. However, this protocol can directly be used to study the effect of different growth conditions, including nutritional stress (e.g., phosphate deficiency), on this process. Importantly, this protocol can readily be adapted for the analysis of other fluorescent compounds that accumulate in the plant root tissues.

2 Materials

2.1 Plant Materials

Arabidopsis thaliana ecotype Columbia (Col-0) as wild type (WT)

Coumarin biosynthesis mutants, *f6′h1-1* [14], *s8h-2* [15], and *cyp82C4-1* [3].

2.2 Plant Growth Medium and Conditions

1. For Fe sufficient or deficient growth conditions: germinate seeds on ½ Murashige and Skoog medium (½MS, pH 5.7) with 0.05% (w/v) MES, 1% sucrose, and 0.7% agar, and grow under long-day conditions (16 h light/8 h dark cycle) in the presence (+Fe) or absence (−Fe) of 50 μM Fe provided as Fe (III)-EDTA.

2. For alkaline growth conditions: germinate seeds on ½ MS (pH 5.7) containing 50 μM Fe(III)-EDTA for 4 days and then transfer for another 3 days of growth to ½ MS medium prepared with 0.5 g/L HEPES buffer, adjusted to pH 7.3 with KOH.

2.3 Reagents

1. Propidium iodide.
2. Scopolin.
3. Fraxin.
4. Esculin.
5. Milli-Q water buffered with MES pH 5.2 or HEPES pH 7 (both at 0.5 g/L).

2.4 Labware and Equipment

1. 1.5 mL microfuge tubes
2. Zeiss microscope equipped with spectral detector.
3. Microscope slide.
4. Cover slip.
5. Forceps.
6. Small Petri dish.
7. Pipette.
8. Tips.

3 Methods

3.1 Plant Growth and Treatment

1. Surface-sterilize and sow Arabidopsis seeds on ½ MS agar medium containing 50 μM Fe (+Fe) or without Fe (−Fe) in 12 × 12 cm square Petri dishes.
2. Put the Petri dishes with Arabidopsis seeds in the growth chamber at 22 °C under long-day conditions (16 h light/8 h dark cycle) for 7 days.
3. For alkaline growth conditions, transfer 4-day-old seedlings to ½ MS agar medium containing 0 μM Fe (Fe deficiency) and pH adjusted to 7.3, and place them back under long-day conditions for another 3 days.

3.2 Spectral Library Setup

1. Prepare Milli-Q water buffered at pH 5.2 and pH 7 using MES and HEPES, respectively (*see* **Note 1**).
2. Dissolve (1 mg) scopolin, fraxin, and esculin in (20 μL) water buffered at pH 5.2 and pH 7. Drop 5 μl of each solution on a microscope slide and place a cover slip.

3. Place the slide with dissolved coumarins on the LSM 880 multiphoton microscope (Zeiss, Oberkochen, Germany) equipped with a Chameleon Ultra II laser (Coherent, Santa Clara, CA, USA) and a W Plan Apochromat 920 1.0 objective. Excitation wavelength should be set at 720 nm using a two photons infrared light, giving a 360-like nm excitation, reproducing UV laser monophotonic excitation. Use a 32-channel GaAsP spectral detector for spectral imaging, with a spectral resolution of 8.9 nm. Set the focus on the liquid drop and acquire the emission spectrum using spectral imaging with the emission window set from 411 to 691 nm (Fig. 1; *see* **Note 2**).

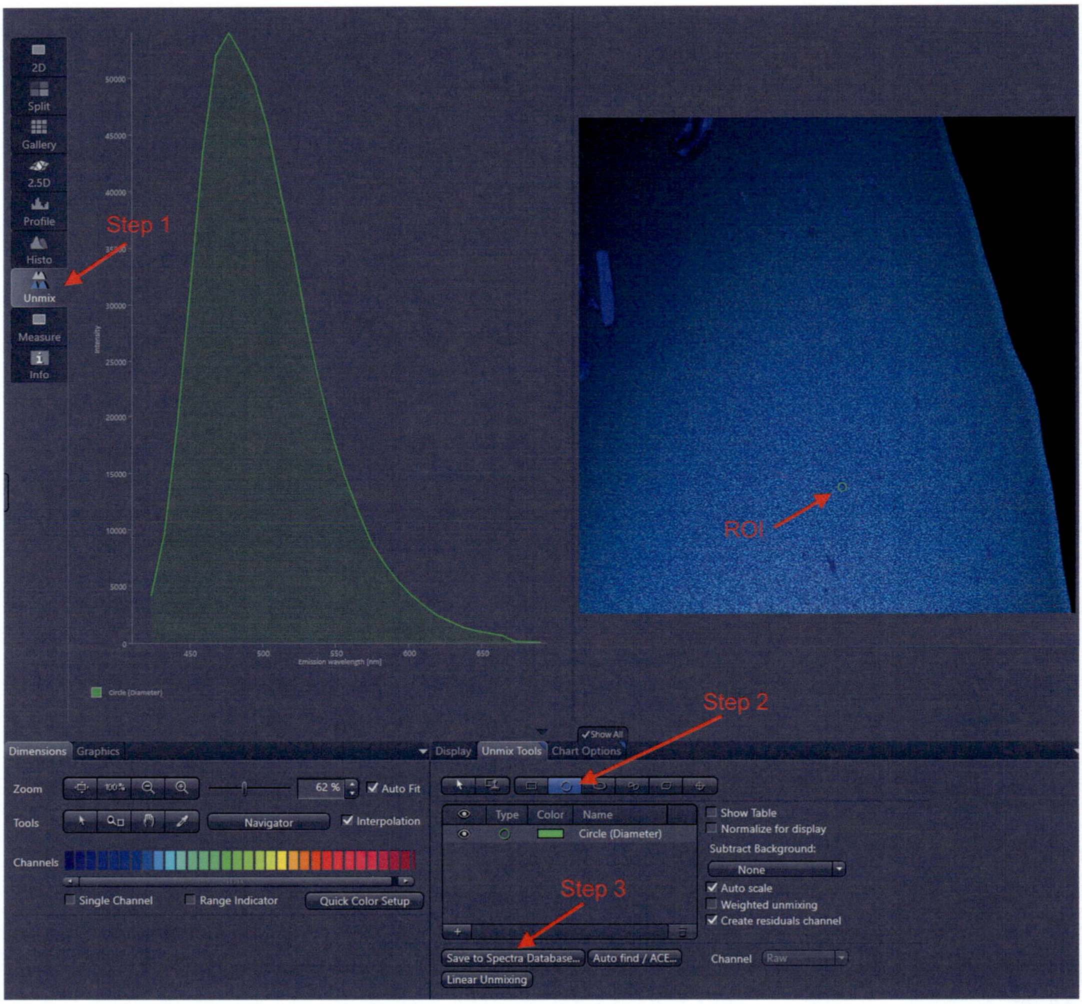

Fig. 1 Spectral library setup. A 5 μL drop containing dissolved coumarins placed on a slide and covered with a cover slip is placed on an LSM 880 multiphoton microscope (Zen). The Zen software is then used to obtain the emission spectra of coumarins as follow: Step 1, Select the unmix function; Step 2, Select a ROI (region of interest) on the pictures and Step 3, save the emission spectrum in the database. Repeat with all the compounds of interest including propidium iodide that labels cell walls

4. Using the unmixing function of the Zen software, draw an ROI (region of interest) on the fluorescent drop and save the emission spectra in the database.
5. Repeat the same procedure for all the studied coumarins.
6. Save the emission spectra of propidium iodide as well to allow visualization of the cell wall after linear unmixing.

3.3 Coumarin Imaging

1. Prepare a solution of propidium iodide (PI, 10 μg/mL) in Milli-Q water in a small Petri dish. Keep the solution in the dark by covering it with aluminum foil.
2. Carefully place a few seedlings into the PI solution to stain the root cell wall and keep it in the dark at room temperature for 10 min. Use seven-day-old seedlings for imaging.
3. Gently rinse the seedlings in Milli-Q water for 1 min and place them on the microscope slide, then gently place a cover slip.
4. Coumarins were imaged with the same microscope used in **step 5**. The emission window was set from 411 to 691 nm.
5. Image coumarins located in the root of WT plants using the above parameters. Acquire several images for each plant.
6. Image coumarins in biosynthesis mutants to ensure the proper identification of individual coumarins by the linear unmixing software (Fig. 2).
7. Save all the pictures.

3.4 Linear Unmixing

1. Select your picture of interest and choose the emission spectra of the coumarins you are interested in localizing.
2. Use the Linear unmixing function on the ZEN v.2.10 software with linear unmixing function to separate, pixel by pixel, the mixed signals of four compounds: scopolin, fraxin, esculin, and PI (Fig. 3; *see* **Note 3**).
3. Save the data.

4 Notes

1. Because pH might influence the emission spectrum of the compounds of interest, it is important to check the emission spectra at different pH. As coumarins are mostly found in the vacuole (~ pH 5) and cytoplasm (~ pH 7), the effect of the two pH was assessed on the emission spectra of scopoline, fraxin, and esculin.
2. Focusing might be a bit difficult due to the thin layer of the solution.

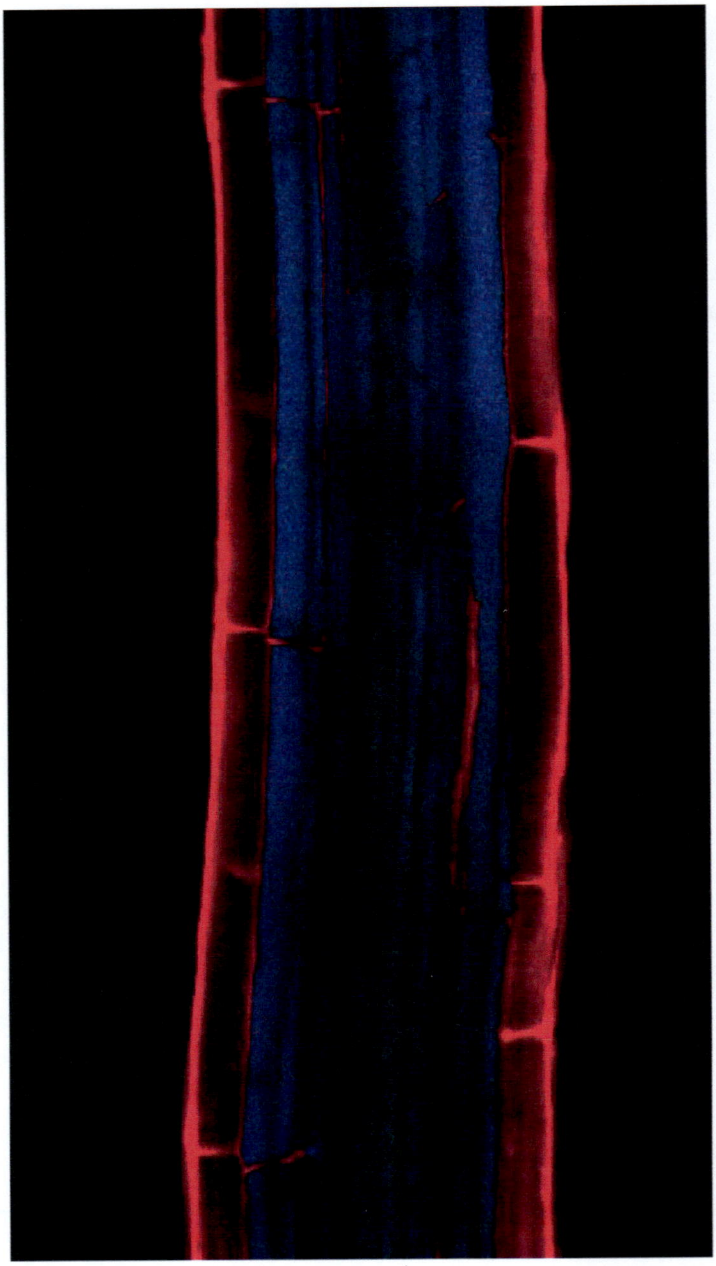

Fig. 2 *In planta* coumarin spectral imaging obtained from wild type root seedlings. Excitation: 720 nm. Emission: 411–690 nm. Red: propidium iodide, blue: coumarins

3. Do not forget to add the PI emission spectra in the library if you want to visualize the cell wall. The residual channel can also be added. The residual channel shows the emission spectra that are not explained by the coumarin spectra library used (Fig. 4).

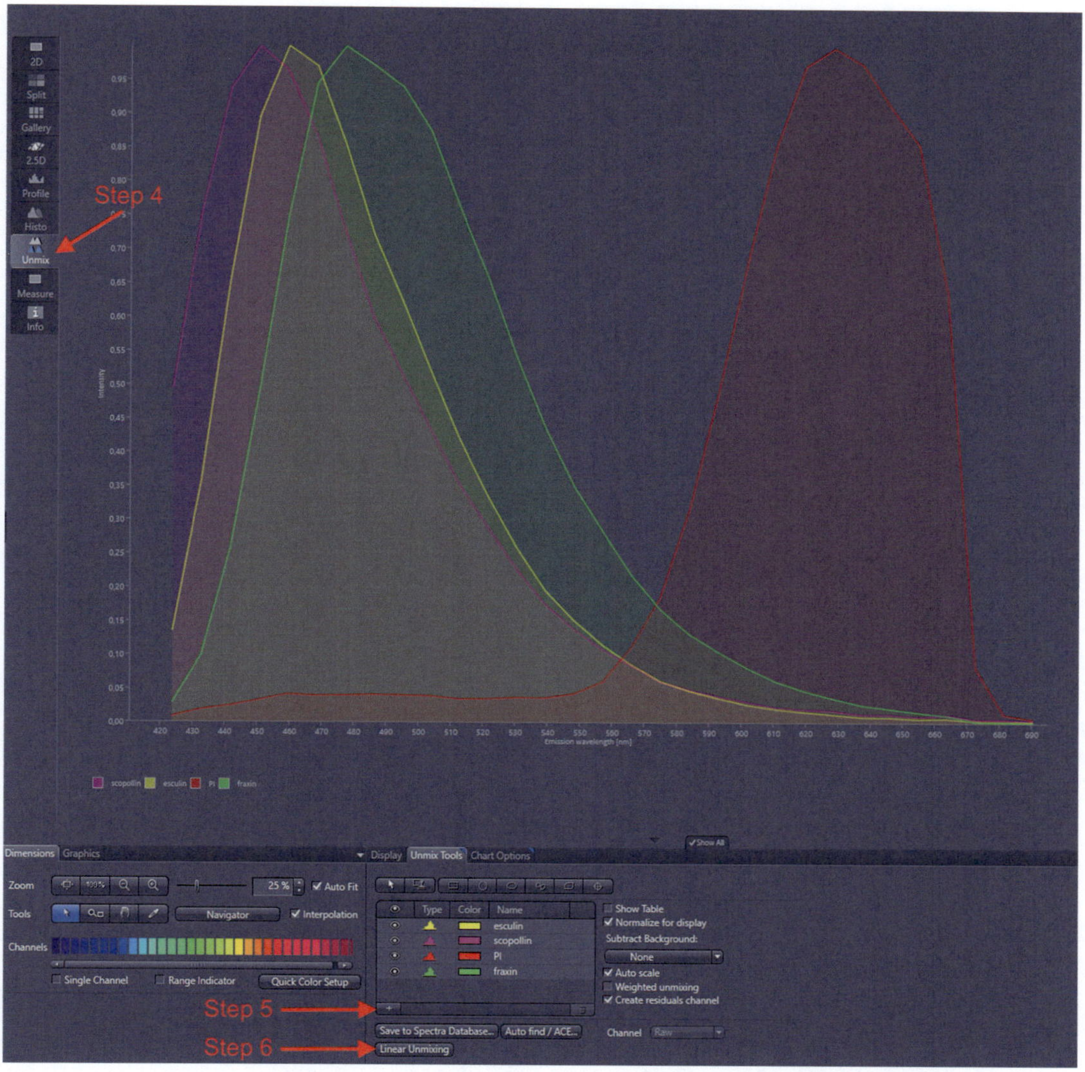

Fig. 3 Linear unmixing. Open the spectral image of a plant root sample and select the unmix function (Step 4). Then click the "+" button (Step 5) to add the different emission spectra of interest that were previously saved (Steps 1–3). Last, Click on linear unmixing button (Step 6). Emission spectra in yellow, green, magenta and red correspond to those of esculin, fraxin, scopolin and propidium iodide

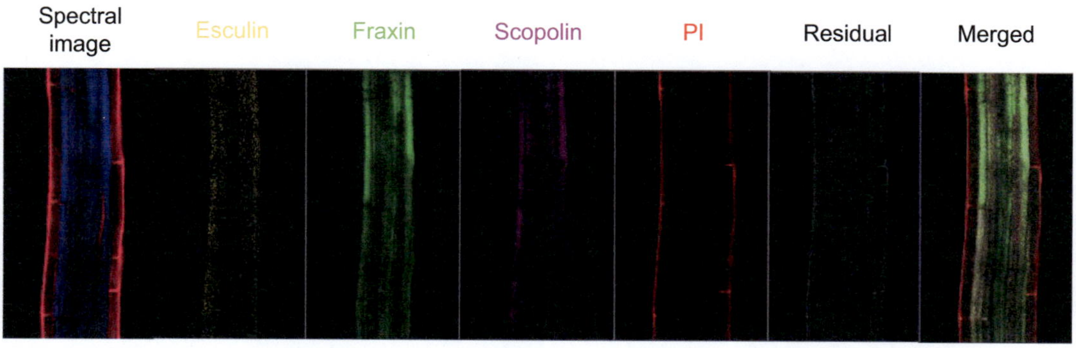

Fig. 4 Linear unmixing of three different coumarins in the root of wild type seedlings grown under alkaline condition. *PI* propridium iodide used to label cell walls

References

1. Gao F, Dubos C (2021) Transcriptional integration of plant responses to iron availability. J Exp Bot 72:2056–2070. https://doi.org/10.1093/jxb/eraa556
2. Robe K, Izquierdo E, Vignols F et al (2020) The coumarins: secondary metabolites playing a primary role in plant nutrition and health. Trends Plant Sci 26:248. https://doi.org/10.1016/j.tplants.2020.10.008
3. Rajniak J, Giehl RFH, Chang E et al (2018) Biosynthesis of redox-active metabolites in response to iron deficiency in plants. Nat Chem Biol 14:442–450. https://doi.org/10.1038/s41589-018-0019-2
4. Rodríguez-Celma J, Schmidt W (2013) Reduction-based iron uptake revisited: on the role of secreted iron-binding compounds. Plant Signal Behav 8:8
5. Tsai H-H, Rodríguez-Celma J, Lan P et al (2018) Scopoletin 8-hydroxylase-mediated Fraxetin production is crucial for iron mobilization. Plant Physiol 177:194–207. https://doi.org/10.1104/pp.18.00178
6. Vanholme R, Sundin L, Seetso KC et al (2019) COSY catalyses trans-cis isomerization and lactonization in the biosynthesis of coumarins. Nat Plants 5:1066. https://doi.org/10.1038/s41477-019-0510-0
7. Fourcroy P, Sisó-Terraza P, Sudre D et al (2014) Involvement of the ABCG37 transporter in secretion of scopoletin and derivatives by Arabidopsis roots in response to iron deficiency. New Phytol 201:155–167. https://doi.org/10.1111/nph.12471
8. Tsai H-H, Schmidt W (2020) pH-dependent transcriptional profile changes in iron-deficient Arabidopsis roots. BMC Genomics 21:694. https://doi.org/10.1186/s12864-020-07116-6
9. Robe K, Conejero G, Gao F et al (2021) Coumarin accumulation and trafficking in Arabidopsis thaliana: a complex and dynamic process. New Phytol 229:2062–2079. https://doi.org/10.1111/nph.17090
10. Zimmermann T, Rietdorf J, Pepperkok R (2003) Spectral imaging and its applications in live cell microscopy. FEBS Lett 546:87–92. https://doi.org/10.1016/s0014-5793(03)00521-0
11. Garini Y, Young IT, McNamara G (2006) Spectral imaging: principles and applications. Cytometry A 69:735–747. https://doi.org/10.1002/cyto.a.20311
12. Koyyappurath S, Conéjéro G, Dijoux JB et al (2015) Differential responses of vanilla accessions to root rot and colonization by fusarium oxysporum f. sp. radicis-vanillae. Front Plant Sci 6:1125. https://doi.org/10.3389/fpls.2015.01125
13. Talamond P, Verdeil J-L, Conéjéro G (2015) Secondary metabolite localization by autofluorescence in living plant cells. Molecules 20:5024–5037. https://doi.org/10.3390/molecules20035024
14. Kai K, Mizutani M, Kawamura N et al (2008) Scopoletin is biosynthesized via ortho-hydroxylation of feruloyl CoA by a 2-oxoglutarate-dependent dioxygenase in Arabidopsis thaliana. Plant J 55:989–999. https://doi.org/10.1111/j.1365-313X.2008.03568.x
15. Siwinska J, Siatkowska K, Olry A et al (2018) Scopoletin 8-hydroxylase: a novel enzyme involved in coumarin biosynthesis and iron-deficiency responses in Arabidopsis. J Exp Bot 69:1735–1748. https://doi.org/10.1093/jxb/ery005

Chapter 3

Assay of Fe(III) Chelate Reductase Activity in *Arabidopsis thaliana* Root

Sun A. Kim

Abstract

A sensitive FerroZine assay is used to measure the membrane-bound ferric-chelate reductase activity in the *Arabidopsis thaliana* roots. In Arabidopsis, *FRO2* (FERRIC CHELATE REDUCTASE 2) encodes the Fe (III) chelate reductase and its expression is induced by iron deficiency. As FRO2 reduces Fe(III) to soluble Fe(II), the resulting Fe(II) forms a purple-colored complex with the dye FerroZine. The concentration of the Fe(II)-FerroZine is directly proportional to the absorbance at 562 nm.

Key words Roots, Iron deficiency, *frd1-1*, *frd3-1*, Fe(III) chelate reductase, Fe(III)-EDTA, Fe(II)-Ferrozine, Microplates, Spectrophotometer

1 Introduction

Plants have developed two mechanisms, reduction-based and chelation-based mechanisms, to mobilize insoluble ferric oxyhydrates (Fe(III) oxide-hydroxide) in the rhizosphere. Arabidopsis thaliana induces a set of biochemical activities required for reduction-based iron uptake in response to iron deficiency [1]. Root plasma membrane H^+-ATPases release protons to acidify the rhizosphere and increase Fe release in the soil [2]. Fe(III) is then reduced to Fe(II) by the membrane-bound ferric-chelate reductase, FERRIC CHELATE REDUCTASE 2 (FRO2) [3]. Finally, the resulting Fe(II) is transported into root epidermal cells by a high-affinity transporter, IRON-REGULATED TRANSPORTER 1 (IRT1) [4].

A colorimetric Ferrozine assay monitors the formation of purple-colored Fe(II)-FerroZine complex and the absorbance of the complex is measured spectrophotometrically [5]. The assay was adopted to screen Arabidopsis mutants that were defective in Fe

Jeeyon Jeong (ed.), *Plant Iron Homeostasis: Methods and Protocols*, Methods in Molecular Biology, vol. 2665,
https://doi.org/10.1007/978-1-0716-3183-6_3,

(III) chelate reductase activity [6]. Fe(III) containing solution was incubated with iron deficient roots and the concentration of Fe(II)-FerroZine was analyzed to quantify the activity of the root Fe(III) chelate reductase.

2 Materials

1. Arabidopsis thaliana accessions.
   Col-*gl1* [ABRC, stock number CS3879]
   frd1-1 [ABRC, stock number CS3777]
   frd3-1 [ABRC, stock number CS6584]
2. Seed surface sterilization solution:
   25% bleach
   0.05% SDS
3. Growth media:
   Half-strength Gamborg's B5 basal salts
   1× Gamborg's vitamin
   1 mM MES
   adjusted to pH 5.8
   0.7% (w/v) type M agar (*see* **Notes 1** and **2**)
4. Iron-sufficient and iron-deficient media:

10× Macronutrients:		
	gram in L	10× Stock conc (mM)
$Ca(NO_3)_2$	4.724 g	20 mM
K_2SO_4	1.307 g	7.5 mM
$MgSO_4$	1.602 g	6.5 mM
KH_2PO_4	0.136 g	1 mM

1000× Micronutrients:		
	gram in L	1000× Stock conc (mM)
H_3BO_3	0.6183 g	10 mM
$MnSO_4$	0.0169 g	0.1 mM
$CuSO_4$	0.0125 g	0.05 mM
$ZnSO_4$	0.014 g	0.05 mM
$(NH_3)_6Mo_7O_{24}$	0.0618 g	0.005 mM

Iron stock solution (50 mM Fe):	
0.9178 g	Fe(III)NaEDTA
413 μL	concentrated HCl
bring to 50 mL with distilled water	
filter sterilize and store in dark at 4 °C	

Iron-sufficient media:
1× Macronutrients
1× Micronutrients
1 mM MES
adjusted to pH 6.0
1% (w/v) type M agar (*see* **Notes 1** and **2**)
After autoclaving, add iron stock solution to make final 50 μM Fe(III)-EDTA

Iron-deficient media:
1× Macronutrients
1× Micronutrients
1 mM MES
adjusted to pH 6.0
1% (w/v) type M agar
After autoclaving, add powder to make final 300 μM Ferrozine

5. Reductase assay solution (*see* **Note 3**):

   100 μM Fe(III)-EDTA

   300 μM FerroZine
6. Tube rocker.
7. Spark® microplate reader, Tecan.
8. 48-well microplate.
9. 96-well microplate, flat bottom (Costar® 96-well, catalog number 3370).
10. Fine forcep or dissecting tweezer.

3 Methods

3.1 Preparation of Arabidopsis Plants

1. Aliquot Arabidopsis seeds in microcentrifuge tubes.
2. In a laminar flow hood, add 1 mL of the seed surface sterilization solution into the microcentrifuge tube with seeds and incubate on a tube rocker at room temperature for 20 min. Rinse with 1 ml sterile water at least 5 times. Add 1 mL sterile water and wrap microcentrifuge tubes with foil.
3. Store sterilized seeds at 4 °C in the dark for 3 days.
4. Sow seeds on the growth media and wrap the plates with surgical tape.
5. Move the plates to a growth incubator with a 16-h light/8-h dark cycle at 22–24 °C. Let plants grow until the 5th/6th true leaves emerge. This typically takes 10–14 days.
6. Carefully pull roots out of the growth media using dissecting tweezer or forceps and transfer seedlings to iron-sufficient or iron-deficient media in square petri dishes. Align roots in parallel on the surface of the media.
7. Move the plates to a growth incubator and keep square petri dishes vertically for 3 days.

3.2 Ferric Reductase Assay

1. Add 500 μL of reductase assay solution to a 48-well plate. Have an extra well with assay solution.
2. Take a plant and submerge the entire root into the individual well. Leave a well without a plant as a reagent blank. Incubate the plate at room temperature in the dark for 30 min.
3. Mix the assay solution using pipettes and transfer 160 μL assay solution to a flat bottom 96-well plate (*see* **Note 4**).
4. Measure the absorbance at 562 nm. Use the assay solution without a plant as a reagent blank.

3.3 Result Presentation

1. Use the molar absorption coefficient of 28.6 mM^{-1} cm^{-1} to calculate the concentration of the Fe(II)-FerroZine complex (*see* **Note 5**).

$$A = \varepsilon c l$$

- A = absorbance.
- ε = the molar absorption coefficient (mM^{-1} cm^{-1})
- c = the molar concentration of the sample (mM),
- l = the pathlength (cm).

$$\text{Concentration of Fe(II)} - \text{FerroZine } (\mu\text{M}) = [\text{absorbance}/0.0286/0.5]$$

2. The amount of reduced iron is adjusted by the volume of the assay solution and the reaction time (*see* **Note 6**). The reductase activity may be normalized by the fresh weight of the root. Excise and blot roots thoroughly on Kimwipe before weighing (*see* **Note 7**). Present the ferric chelate reductase activity in the unit of μmol Fe(II) $\text{gram}^{-1}\ \text{h}^{-1}$.

$$\text{Reductase activity } \left(\mu\text{mol Fe(II) gram}^{-1}\ \text{h}^{-1}\right) = [\text{absorbance}/0.0286/0.5]^{*}\ 0.0005\ (\text{L})/\text{root weight (gram)}\ /0.5\ (\text{hour})$$

3. The results are presented in the unit of μmol Fe(II) $\text{seedling}^{-1}\ \text{h}^{-1}$ (*see* **Note 8**).

4 Notes

1. Lowering the percentage of agar to 0.6% may prevent root breakage during the transfer at the following step.
2. Prepare media in square Petri dishes.
3. The reductase assay solution needs to be made fresh every time before the assay. Weight out powder each time and keep the solution in the dark during the preparation.
4. Fe deficiency will increase the formation of the purple-colored Fe(II)-ferrozine complex around the roots of wild-type (Col-gl1) plants. *frd1-1* mutants do not induce ferric reductase activity; *frd3-1* mutants constitutively induce ferric reductase activity under both Fe conditions.
5. The pathlength is 1 cm for a conventional spectrophotometer. The pathlength is 1 mm (without auto-ranging to 0.2 mm) if measured by NanoDrop spectrophotometer. The light pathlength is 5 mm for 160 μL assay solution in a flat bottom 96-well. Consult the user manual and verify the pathlength.
6. The volume of assay solution in **step 1** (500 μL) and the reaction time in **step 2** (30 min).
7. To measure the fresh weight of roots more accurately, you may pool 3–5 seedlings as a group.
8. The seedlings can be transferred to soil to continue to grow the plants after the assay.

References

1. Kim SA, Guerinot ML (2007) Mining iron: iron uptake and transport in plants. FEBS Lett 581: 2273–2280
2. Santi S, Schmidt W (2009) Dissecting iron deficiency-induced proton extrusion in Arabidopsis roots. New Phytol 183:1072–1084
3. Robinson NJ, Procter CM, Connolly EL et al (1999) A ferric-chelate reductase for iron uptake from soils. Nature 397:694–697
4. Eide D, Broderius M, Fett J et al (1996) A novel iron-regulated metal transporter from plants identified by functional expression in yeast. Proc Natl Acad Sci U S A 93:5624–5628
5. Gibbs CR (1976) Characterization and application of FerroZine iron reagent as a ferrous iron indicator. Anal Chem 48:1197–1201
6. Yi Y, Guerinot ML (1996) Genetic evidence that induction of root Fe(III) chelate reductase activity is necessary for iron uptake under iron deficiency. Plant J 10:835–844

Chapter 4

An Adapted Protocol for Quantitative Rhizosphere Acidification Assay

Sara Omer, Claire Macero, Dayishaa Daga, Kelly Zheng, and Jeeyon Jeong

Abstract

Acidification of the rhizosphere is a key process in the homeostasis of multiple essential nutrients, including iron. Under iron deficiency, the release of protons from the roots helps solubilize and increase the accessibility of iron in the soil. Rhizosphere acidification has been widely examined in many iron homeostasis studies, generally using a qualitative method based on the color change of bromocresol purple, a pH indicator dye, near the roots. In this chapter, we introduce an adapted version of a rhizosphere acidification assay protocol that allows for the quantitative assessment of small pH changes in the rhizosphere. This colorimetric method also utilizes bromocresol purple, but the ratio of its absorbance at 434 nm and 588 nm is considered to quantify protons released into the assay solution. Furthermore, the assay is compatible with small sample volumes, such as those with young Arabidopsis seedlings.

Key words Iron, Rhizosphere acidification, Bromocresol purple, Colorimetric, Proton

1 Introduction

Rhizosphere acidification plays a key role in nutrient homeostasis. Increased proton efflux from the roots has been observed under deficiencies of essential nutrients such as nitrogen, phosphorus, potassium, and iron, and contributes to enhancing the accessibility of the nutrients in the soil solution [1, 2]. Under iron-deficient conditions, dicots use a reduction-based mechanism that involves rhizosphere acidification. This process, referred to as Strategy I, is induced by iron deficiency and consists of the extrusion of protons by proton ATPases on the plasma membrane of root epidermal cells [3], the reduction of ferric chelates to ferrous iron by Ferric Reductase Oxidase 2 (FRO2) [4], secretion of coumarins that bind and mobilize iron [5], and the uptake of ferrous iron by the high-affinity iron transporter, Iron Regulated Transporter 1 (IRT1) [6–8]. The reduced pH around the roots helps solubilize ferric chelates and increases iron availability [3]. While rhizosphere

Jeeyon Jeong (ed.), *Plant Iron Homeostasis: Methods and Protocols*, Methods in Molecular Biology, vol. 2665,
https://doi.org/10.1007/978-1-0716-3183-6_4,

acidification is assumed as the first step in Strategy I, it has been proposed that proton extrusion is a response induced at later stages of iron deficiency that accompanies alterations in root development to seek additional iron available in the soil more effectively [3].

As a major process of Strategy I, iron deficiency-induced rhizosphere acidification has been widely examined in iron homeostasis research, including landmark studies that investigated shoot-to-root iron signaling [9], the *frd1* mutant [10] that is allelic to *FRO2* [4], *FRD3* [11], *POPEYE* [12], and *URI* [13]. The most commonly used method to assess rhizosphere acidification employs a solid medium containing bromocresol purple, a pH indicator dye that enables the observation of acidification surrounding the roots with the naked eye based on changes in the medium color [14]. Bromocresol purple emits in the visible range and results in a vivid color change that is caused by pH-dependent transitions between two degenerate intramolecular charge transfer states [15]. Bromocresol purple is a reliable indicator of pH changes between approximately 5.8 and 7.5 [16, 17], which is the appropriate range to study rhizosphere acidification. In acidic pH, bromocresol purple molecules lose electrons and exhibit light yellow color, whereas, upon pH increase, more BCP molecules are negatively charged and purple. While the assay on solid medium containing bromocresol purple is convenient, it does not allow for quantification of pH changes in the rhizosphere. For a quantitative assessment of rhizosphere acidification, the pH of the assay solution can be directly measured, an approach that has been used in the early studies on rhizosphere acidification with bean plants [18] and Arabidopsis plants grown in liquid medium [19, 20]. However, when working with young Arabidopsis seedlings, accurately measuring changes in the pH of a small volume of assay solution can be technically challenging.

Here, we report an adapted version of a rhizosphere acidification assay that utilizes bromocresol purple and quantitatively measure proton levels in the rhizosphere in small assay volumes. We modified the solid medium-based method to a colorimetric liquid assay that involves measuring the color change of bromocresol purple from a small aliquot of assay solution using a 96-well plate spectrophotometer. Absorptions of bromocresol purple at 588 nm and 434 nm are measured, and the ratio of the absorbance at these wavelengths is used to calculate changes in the proton levels. The specific wavelengths were chosen based on the maximum absorbance of deprotonated yellow bromocresol purple solution at 434 nm and that of the protonated violet BCP solution at 588 nm [21]. Similar assays based on the absorbance of bromocresol purple at 590 nm have been reported [3, 22], but accounting for absorbances at both wavelengths helps sensitively quantify small changes in pH. In this study, we verified a strong correlation between the ratio of absorbances at the two wavelengths and

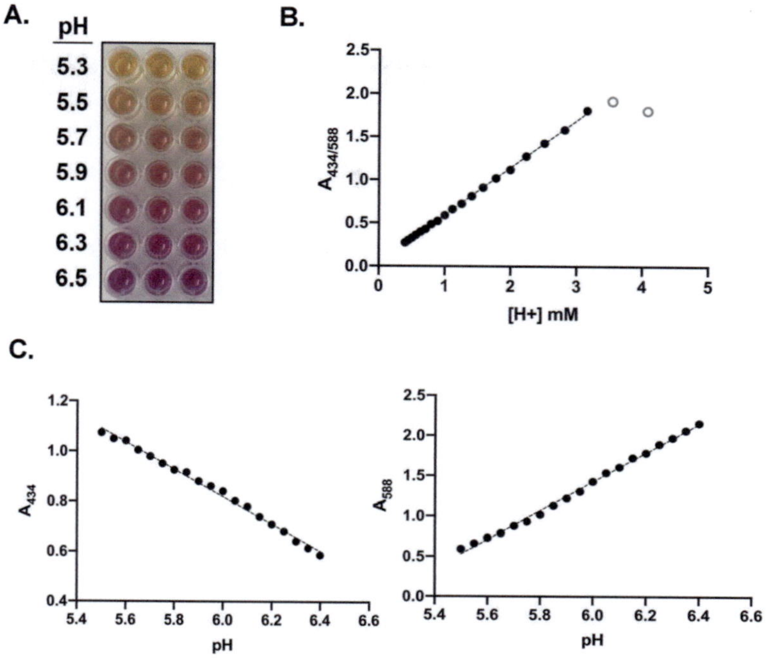

Fig. 1 Standard curve. (a). Representative image of a 96-well plate containing pH standards in the assay buffer in triplicates. (**b**). A standard curve plotted with standard solutions ranging from pH 5.4 to 6.4 in 0.05 increments. The filled circles (●) represent data points within the linear region. $R^2 = 0.9994$, $y = 0.5498x + 0.04323$. (**c**). Absorbance of bromocresol purple at 434 nm (*left*) or 588 nm (*right*) over pH. A_{434}: $R^2 = 0.9925$, $y = -0.5428x + 4.076$, A_{588}: $R^2 = 0.9990$, $y = 1.796x - 9.35$

change in pH (Fig. 1), which was consistently observed in a previous biochemical study [21]. Given that extrusion of protons across the plasma membrane is a key process involved in the acquisition of many essential nutrients [1, 2], this method can be broadly applied in plant biology research.

2 Materials

2.1 Plant Materials and Media

2.1.1 Arabidopsis thaliana *Wildtype Seeds*

See (**Note 1**).

2.1.2 Murashige and Skoog (MS) Medium Without Sucrose (for 500 mL)

- 2.15 g of Murashige & Skoog basal salts with FeNaEDTA (Caisson, MSP34)
- 0.25 g MES (0.05% w/v)

- Adjust pH to 5.8 using KOH.
- 2.75 g agar (0.55% w/v).

2.1.3 Media With or Without Iron (See Note 2)

Iron-Sufficient (+Fe) medium (for 500 mL)

- 2.15 g Murashige & Skoog basal salts with FeNaEDTA (Caisson, MSP34)
- 0.05% w/v MES (0.05% w/v)
- Adjust pH to 6.0 using KOH.
- 0.55% w/v agar.

Iron-Deficient (−Fe) Medium (for 500 mL)

- 2.135 g Murashige & Skoog basal salts without iron (Caisson, MSP33)
- 0.05% w/v MES
- Adjust pH to 6.0 using KOH
- 0.55% w/v agar
- 310.6 μM FerroZine, to be added after autoclaving.

2.2 Solutions and Reagents

- Assay solution (425 mL): 0.2 mM calcium sulfate, adjust pH to 6.5 with NaOH.
- pH standard solutions (approximately 80 ml each): 0.2 mM calcium sulfate, 0.05% w/v MES (*see* **Note 3**). 0.04% Bromocresol purple stock solution.
- Seed surface sterilization solution: 70% v/v ethanol, 0.05% v/v Triton-X100.
- 0.1 M and 1 M NaOH for pH calibration.

2.3 Equipment

- Microplate Spectrophotometer.
- 96-well spectrophotometer plate
- Microbalance.
- Biosafety cabinet.
- Plant growth chamber.
- pH meter,
- Micropore tape.
- 24-well culture plate
- Kimwipes.
- Tweezers.
- Petri dishes.
- Razor blade or dissecting scissors.
- Multichannel pipette.
- Timer.

3 Methods

3.1 Preparation of Plant Samples

1. Surface sterilize seeds and sow on MS medium without sucrose (*see* **Note 4**).
2. Secure the lids of the plates with micropore tape ensuring that all exposed regions are taped.
3. Place the plates vertically in a growth chamber and grow the plants for 7–10 days.
4. Three days prior to the assay, transfer seedlings using a tweezer to iron-sufficient and -deficient medium (*see* **Note 5**).
5. Seal the closed plates with micropore tape and return them to the vertical position in the growth chamber for 72 h.

3.2 Preparation of Assay Solution

1. Dissolve 0.0172 g calcium sulfate dihydrate in approximately 420 mL of Milli-Q water in a beaker. The final concentration of calcium sulfate in 425 mL will be 0.2 mM.
2. Adjust the pH to 6.5 with 0.1 M sodium hydroxide (*see* **Note 6**).
3. Bring the final volume to 425 mL with Milli-Q water.
4. Transfer the solution to a clean and sterile glass bottle (*see* **Note 7**).

3.3 Preparation of pH Standard Solutions

1. Prepare approximately 80 mL of a solution of 0.2 mM calcium sulfate and 0.05% MES w/v.
2. Adjust the pH to 5.3 with 1 M sodium hydroxide.
3. Repeat these steps to prepare additional pH standards (*see* **Note 3**).

3.4 Incubation of Samples in Assay Solution

1. Aliquot ~25 mL of the assay solution into two petri dishes: each aliquot will be used for samples grown on +Fe or −Fe medium (*see* **Note 8**).
2. Make an approximately 1.5 mL aliquot of the assay solution in a microcentrifuge tube (*see* **Note 9**).
3. Pick up seedlings one by one, gently dap on Kimwipe to remove agar stuck to the roots, and rinse the root in the petri dish with aliquoted assay medium. Stack seedlings on top of each other on the inside of the cover of the petri dish to keep roots and shoots separated (*see* **Note 10**).
4. Once stacked, gently dab on Kimwipe and carefully place the roots into the desired well in a 24-well plate. Make sure the roots do not dry out.
5. Add 2 mL of the assay solution to each well and cover the plate to avoid evaporation (*see* **Note 11**). Immediately move plates with samples and the assay solution aliquot from **step 2** to the growth chamber.

6. Set a timer to keep track of time for the time points.
7. For time point zero, transfer 170 μL of assay solution to a 96-well plate and add 30 μL of 0.04% BCP for a final working concentration of 0.006% BCP. Prepare in triplicates.
8. Take absorbance of the blank assay solution with BCP for time point zero.
9. Proceed to making the standard curve.

3.5 Generation of a Standard Curve

1. After the samples are placed in the growth chamber and the absorbances at time point zero have been measured, aliquot 170 μL of each pH standard solution to three wells in a 96-well plate.
2. Add 30 μL of 0.04% BCP to each well using a multi-channel pipette. Ensure BCP is mixed evenly with the pH standard solution. Incubate at room temperature for 1 min. Solution color should change from transparent to the color corresponding to the pH level (*see* **Note 12**; Fig. 1a).
3. Measure the absorbance of each well at 434 nm and 588 nm.
4. Calculate the average absorbance at 434 nm (A_{434}) and 588 nm (A_{588}) for each triplicate pH standard at each wavelength.
5. Using the average absorbances, calculate the ratio of A_{588} and A_{434} for each pH standard to obtain $A_{434/588}$.
6. Plot the proton concentration on the x-axis and the $A_{434/588}$ in the y-axis (*see* **Note 13**; Fig. 1b).
7. Fit a linear regression line to the graph and record the line equation (*see* **Note 14**).

3.6 Measurement of Absorbance and Root Fresh Mass

1. Remove the samples from the growth chamber and transfer 170 μL of the assay solution from each well to a corresponding well in the 96-well plate (*see* **Note 15**).
2. Add 30 μL of the BCP stock solution and ensure BCP is mixed well with the assay solution by gently pipetting up and down 3–4 times.
3. Incubate at room temperature for 1 min.
4. Measure the absorbance of each well at 434 nm and 588 nm. Measure in triplicates and record values.
5. After the last time point, add Milli-Q water to a petri dish and rinse the roots by dipping them into the water (*see* **Note 16**). Dab on Kimwipe to remove excess liquid.
6. Lay the seedlings on a petri dish and use a razor blade or dissecting scissors to cut the shoots off the roots. Using a microbalance, measure the fresh mass of roots (*see* **Note 17**). Record data.

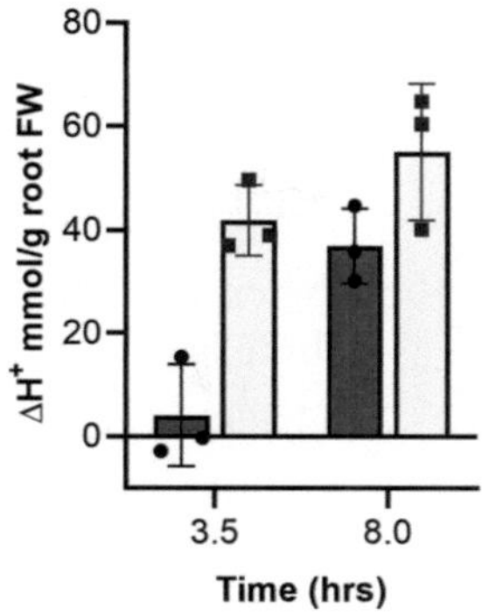

Fig. 2 Sample data. Proton secretion into the assay solution was quantified at time points, 3.5 and 8 h. Each data point represents results obtained from a pooled sample of 25–30 seedlings grown under iron-sufficient (+Fe; ●, grey bars) or iron-deficient (−Fe; ■, white bars) conditions

3.7 Normalization of Sample Data

1. Calculate the average absorbance values at 434 nm (A_{434}) and 588 nm (A_{588}) for each sample at different time points.
2. Using the average absorbance values, obtain the $A_{434/588}$ for each sample at each time point.
3. Calculate the measured proton concentration in each sample at each time point based on the linear fitted equation from the standard curve.
4. To obtain the change in proton concentration due to proton release from the samples, subtract the proton concentration of the blank control from the proton concentration of the sample at each time point.
5. Convert the change in proton concentration from step 4 to moles of protons (*see* **Note 18**).
6. Normalize data by dividing the moles of protons by fresh root mass of each sample.
7. Plot the moles of protons per root fresh mass over different time points (Fig. 2).

4 Notes

1. Depending on the experiment, seeds of other genotypes/ecotypes of interest will be grown alongside their controls.
2. Iron-sufficient and -deficient media can be replaced with alternative media depending on the experimental goals.
3. In this example, pH standards ranging from pH 5.4 to 6.4 in 0.05 increments were used to generate the standard curve. The specific pHs and total number of standards can be adjusted depending on the experiment.

4. Plate 15–20 seeds per plate with even spacing to ensure proper growth.
5. Be careful not to stress the seedlings by damaging the roots. Ensure the root is in contact with the media after transfer.
6. When adjusting the pH of the assay solution, be aware that there is no buffer. The pH will rapidly change.
7. The assay solution and pH standard solution can be kept at 4 ° C, but should be brought to room temperature prior to use.
8. Before starting, check the pH of the assay solution to ensure that it is at 6.5 pH.
9. Keep the microcentrifuge tube with the assay solution next to your samples in the growth chamber. This will be used as a blank control for each time point.
10. Ensuring that the roots and shoots are separated now will keep them from getting tangled throughout the assay. If they are tangled, separating them becomes difficult later on.
11. The 2 mL of assay solution will allow for a maximum of 5 time points in triplicates. Plan accordingly.
12. When working with BCP, keep in mind that yellow corresponds to a lower pH while purple corresponds to a higher pH. A color change should be visible when BCP is added to the pH standards and should follow this trend.
13. Convert pH values to proton concentrations using the equation $pH = -\log [H^+]$.
14. The absorbance ratio of $A_{434/588}$ and proton concentration exhibit a strong linear correlation (Fig. 1b). The absorbances measured at 434 nm or 588 nm each show strong linear correlation with the pH standards (Fig. 1c).
15. Before taking an aliquot from the assay sample, make sure to gently swirl the plate to mix the solution in the wells. This is to ensure that the protons released from the roots are evenly distributed when an aliquot is taken. Keep the seedlings together with tweezers but avoid tangling between the roots and shoots.
16. To avoid dehydration of the seedlings and the resulting loss of fresh weight, immediately weigh the samples.
17. Divide the concentration of protons by the volume of assay buffer when the sample was taken for absorbance measurements. Make sure to consider the volume removed at each time point.

Acknowledgments

This work was supported by the Gregory Call Student Research Fund to SO, CM, and KZ, and the National Science Foundation grants (IOS#1754969 and IOS#2143478) to JJ.

References

1. Houmani H, Rabhi M, Abdelly C, Debez A (2015) Implication of rhizosphere acidification in nutrient uptake by plants: cases of potassium (K), phosphorus (P), and iron (fe). In: Hakeem KR (ed) Crop production and global environmental issues. Springer, Cham, pp 103–122
2. Wang X, Tang C (2018) The role of rhizosphere pH in regulating the rhizosphere priming effect and implications for the availability of soil-derived nitrogen to plants. Ann Bot 121:143–151. https://doi.org/10.1093/aob/mcx138
3. Santi S, Schmidt W (2009) Dissecting iron deficiency-induced proton extrusion in Arabidopsis roots. New Phytol 183:1072–1084. https://doi.org/10.1111/j.1469-8137.2009.02908.x
4. Robinson NJ, Procter CM, Connolly EL, Guerinot ML (1999) A ferric-chelate reductase for iron uptake from soils. Nature 397:694–697. https://doi.org/10.1038/17800
5. Robe K, Izquierdo E, Vignols F et al (2021) The coumarins: secondary metabolites playing a primary role in plant nutrition and health. Trends Plant Sci 26:248–259. https://doi.org/10.1016/j.tplants.2020.10.008
6. Varotto C, Maiwald D, Pesaresi P et al (2002) The metal ion transporter IRT1 is necessary for iron homeostasis and efficient photosynthesis in *Arabidopsis thaliana*. Plant J 31:589–599. https://doi.org/10.1046/j.1365-313X.2002.01381.x
7. Vert G, Grotz N, Dédaldéchamp F et al (2002) IRT1, an Arabidopsis transporter essential for iron uptake from the soil and for plant growth. Plant Cell 14:1223–1233. https://doi.org/10.1105/tpc.001388
8. Connolly EL, Fett JP, Guerinot ML (2002) Expression of the IRT1 metal transporter is controlled by metals at the levels of transcript and protein accumulation. Plant Cell 14:1347–1357. https://doi.org/10.1105/tpc.001263
9. Grusak MA, Pezeshgi S (1996) Shoot-to-root signal transmission regulates root Fe(III) reductase activity in the dgl mutant of pea. Plant Physiol 110:329–334. https://doi.org/10.1104/pp.110.1.329
10. Yi Y, Guerinot ML (1996) Genetic evidence that induction of root Fe(III) chelate reductase activity is necessary for iron uptake under iron deficiency. Plant J 10:835–844. https://doi.org/10.1046/j.1365-313X.1996.10050835.x
11. Rogers EE, Guerinot ML (2002) FRD3, a member of the multidrug and toxin efflux family, controls iron deficiency responses in Arabidopsis. Plant Cell 14:1787–1799. https://doi.org/10.1105/tpc.001495
12. Long TA, Tsukagoshi H, Busch W et al (2010) The bHLH transcription factor POPEYE regulates response to iron deficiency in Arabidopsis roots. Plant Cell 22:2219–2236. https://doi.org/10.1105/tpc.110.074096
13. Kim SA, LaCroix IS, Gerber SA, Guerinot ML (2019) The iron deficiency response in Arabidopsis thaliana requires the phosphorylated transcription factor URI. Proc Natl Acad Sci U S A 116:24933–24942. https://doi.org/10.1073/pnas.1916892116
14. Yi Y, Saleeba JA, Guerinot ML (1994) Iron uptake in *Arabidopsis thaliana*. In: Manthey J, Luster D, Crowley DE (eds) Biochemistry of metal micronutrients in the rhizosphere. Lewis Publishers, Boca Raton, pp 295–307
15. Talone CJ, Gao J, Lynch JR et al (2016) Determination of the ground- and excited-state dipole moments of bromocresol purple in protic and aprotic solvents. Spectrochim Acta A Mol Biomol Spectrosc 156:138–142. https://doi.org/10.1016/j.saa.2015.11.034
16. Lvov Y, Antipov AA, Mamedov A et al (2001) Urease encapsulation in nanoorganized microshells. Nano Lett 1:125–128. https://doi.org/10.1021/nl0100015
17. Schussel LJ, Atwater JE (1995) A urease bioreactor for water reclamation aboard manned spacecraft. Chemosphere 30:985–994. https://doi.org/10.1016/0045-6535(94)00453-2
18. Ric de Vos C, Lubberding HJ, Bienfait HF (1986) Rhizosphere acidification as a response to iron deficiency in bean plants on JSTOR. Plant Physiol 81:842–846

19. Pizzio GA, Paez-Valencia J, Khadilkar AS et al (2015) Arabidopsis type I proton-pumping pyrophosphatase expresses strongly in phloem, where it is required for pyrophosphate metabolism and photosynthate partitioning. Plant Physiol 167:1541–1553. https://doi.org/10.1104/pp.114.254342
20. Rhizosphere Acidification Assay' ' – BIO-PROTOCOL. https://bio-protocol.org/e1676. Accessed 16 May 2019
21. Pei K, Xiong Y, Li X et al (2018) Colorimetric ELISA with an acid–base indicator for sensitive detection of ochratoxin A in corn samples. Anal Methods 10:30–36. https://doi.org/10.1039/C7AY01959A
22. Vengavasi K, Pandey R (2016) Root acidification, a rapid method of screening soybean genotypes for low-phosphorus stress. Ind J Genet Plant Breed 76:213. https://doi.org/10.5958/0975-6906.2016.00025.0

Chapter 5

Techniques to Study Common Root Responses to Beneficial Microbes and Iron Deficiency

Shu-Hua Hsu, Max J. J. Stassen, Corné M. J. Pieterse, and Ioannis A. Stringlis

Abstract

Iron (Fe) plays a central role in the vital processes of a plant. The Fe status of a plant influences growth and immunity, but it also dictates interactions of roots with soil microbiota through the production of Fe mobilizing, antimicrobial fluorescent phenolic compounds called coumarins. To adapt to low Fe availability in the soil, plants deploy an efficient Fe deficiency response. Interestingly, this Fe deficiency response is hijacked by root-colonizing microbes in the root microbiome to establish a mutually beneficial relationship. In this chapter, we describe how we cultivate plants and microbes to study the interaction between plants, beneficial rhizobacteria, and the plant's Fe deficiency response. We describe (a) how we study activity and localization of these responses by assessing gene-specific promoter activities using GUS assays, (b) how we visualize root-secreted coumarins in response to Fe deficiency and colonization by beneficial rhizobacteria, and (c) how we prepare our samples for metabolite extraction and reverse-transcriptase quantitative PCR to analyze the expression of marker genes.

Key words Iron deficiency, Coumarins, *Pseudomonas simiae* WCS417r, Induced systemic resistance, Plant-microbes interaction

1 Introduction

Iron (Fe) is a redox catalyst for essential cellular processes in almost every organism. Although it is an abundant element in the Earth's crust, its bioavailability is hampered by the fact that it mostly occurs as ferric oxide, which is poorly soluble at neutral and high pH [1–3]. Plant immunity and Fe status are tightly linked: plants can use Fe withholding strategies as well as Fe accumulation to combat infectious pathogens, and it has been shown that Fe deficiency directly influences plant resistance [3–6]. Indirectly, Fe status also mediates beneficial interactions with the microbial community in

Authors Shu-Hua Hsu and Max J. J. Stassen have equally contributed to this chapter.

Jeeyon Jeong (ed.), *Plant Iron Homeostasis: Methods and Protocols*, Methods in Molecular Biology, vol. 2665, https://doi.org/10.1007/978-1-0716-3183-6_5,

the rhizosphere. A well-known example is the plant-protective effect of *Pseudomonads* in disease-suppressive soils, where bacterially produced Fe-chelating siderophores limit Fe availability and therewith pathogen growth [7]. Some of these *Pseudomonas* strains can also directly stimulate plant broad-spectrum foliar resistance when colonizing its roots [8]. This response is termed induced systemic resistance (ISR) and has been described in many species, but has been mainly studied in the *Arabidopsis thaliana* (Arabidopsis) and *Pseudomonas simiae* WCS417 (WCS417) model system [8]. This ISR response is known to be dependent on the hormones jasmonic acid and ethylene, but how root colonization translates to shoot resistance and associated long-distance signals has remained elusive [9, 10].

Aside from stimulating plant defense and the production of siderophores, WCS417 also activates parts of the canonical Fe deficiency response in plants [11, 12]. This involves induction by WCS417 of many genes upregulated in roots following Fe deficiency, such as Fe transporter gene *IRT1* and the ferric chelate reductase gene *FRO2* [12]. Part of this Fe-deficiency module that is activated in the roots, namely the transcription factor MYB72 and the β-glucosidase BGLU42, were shown to be integral to the induced resistance response described above [13, 14]. MYB72 and BGLU42 participate in the roots in the production and secretion of coumarins, fluorescent phenolic compounds that have been extensively studied in the last 10 years for their role in Fe acquisition, especially in alkaline soils [14–18]. As such, these compounds might act downstream of MYB72 and BGLU42 as long-distance signals that participate in the induction of foliar defenses [10].

Next to their potential role in ISR, these Fe deficiency-induced compounds can affect plant-microbiota interactions in the rhizosphere [17, 19, 20]. Arabidopsis mutants for the key enzyme in coumarin biosynthesis F6'H1 assembled distinct microbiomes on their roots compared to wild-type Arabidopsis [17, 20]. Coumarins can shape the root-associated microbiota and display selective antimicrobial activity, evidenced by beneficial bacteria such as WCS417 being resistant to their antimicrobial effects, while some soil-borne pathogens are sensitive to them [17]. All in all, the plant's Fe-deficiency response plays a significant role in microbially induced resistance and the direct interaction with beneficial microbes. Here, we describe the protocols that our lab routinely uses to describe and compare different aspects of the plant's response to Fe-deficiency as well as to inoculation with beneficial rhizobacteria. Firstly, we provide a detailed explanation of our plant and bacterial culturing methods, and describe how we prepare tissue for both gene expression as well as targeted metabolomics analysis. Then, we explain how we utilize qPCR and GUS staining to quantify and localize the expression of Fe deficiency and ISR-related genes in response to both WCS417 and Fe deficiency, and during different time periods. We describe the different setups we use to visualize coumarins in liquid media as well as in an agar plate system. These methods provide reproducible results that give us insight into how beneficial bacteria interact with plant physiology, and how this can result in adaptive plant responses.

2 Materials

2.1 Equipment

- Flow cabinets.
- 100–1000 μL (P1000), 20–200 μL (P200), and 2–20 μL (P20) pipettes and corresponding pipette tips.
- Square Petri dishes (120 × 120 mm).
- Petri dishes (94 mm diameter).
- pH meter.
- Tweezers.
- Spreader.
- Parafilm M.
- Bunsen burner.
- VAPOUR-Line Lite Autoclave.
- 96% ethanol.
- Demineralized water.
- Cutting blade (Fisherbrand™ Razor Blades).
- Bio Dancer Benchtop Shaker (New Brunswick).
- Centrifuge that can hold 50-mL Falcon tubes.
- 50-mL Falcon tubes.
- Glass beaker (250 mL).
- 1.5-mL microcentrifuge tubes.
- Plastic containers.
- Fume hood.
- Desiccator.
- CELLSTAR® 12-well multiwell culture plates.
- CELLSTAR® 96-well flat bottom plates.
- Glass beads.
- Vortex mixer.
- TissueLyser II (Qiagen).
- Intelli-Mixer™ RM-2 M (ELMI).
- NanoDrop™ 2000/2000c (Thermo Fisher).
- ViiA7 real-time PCR system (Thermo Fisher).

- S1000™ Thermal Cycle (Bio-Rad).
- Incubators (65 °C, 28 °C, 37 °C).
- Multichannel Pipette (5–10 μL, 200 μL).
- Strip tubes for PCR and cap strips for PCR tubes.
- Microcentrifuge.
- Plate centrifuge.
- QPCR plate (384-Well 0.02 mL) and optical adhesive film.
- Filter Tip, 10 μL.
- Glass microscope slides.
- Glass cover slips.
- UV Transilluminator.
- Plant growth chamber (22 °C; day/night photoperiod: 10 h/14 h; 100 μmol · m^{-2} · s^{-1} light intensity; 70% humidity).
- Plate reader with a fluorescence detector.

2.2 Buffers, Media, and Solutions

2.2.1 Plant Cultivation

1. Seeds of *Arabidopsis thaliana* (accession Col-0) or Arabidopsis *pGENE::GUS* lines (in the Col-0 background) of your gene of interest. In our case, we used GUS lines of coumarin biosynthesis-related genes such as *MYB72* and *F6'H1*.
2. Bleach.
3. 37% (v/v) hydrochloric acid (HCl) in demineralized water.
4. *Murashige and Skoog* [21] *nutrient solution*: 0.44% (w/v) Murashige and Skoog medium including vitamins, 4.7 mM MES Monohydrate 0.5% (w/v) sucrose per liter of demineralized water, supplemented with 10 g of Plant agar. Adjust the pH to 5.5 using 1M KOH.
5. *Hoagland* [22] *nutrient solution*: 2 mM $Ca(NO_3)_2$, 5 mM KNO_3, 2 mM $MgSO_4$, 2.5 mM KH_2PO_2, 70 μM H_3BO_3, 14 μM $MnCl_2$, 1 μM $ZnSO_4$, 0.5 μM $CuSO_4$, 10 μM NaCl, Na_2MoO_4, 4.7 mM MES monohydrate, 0.5% (w/v) sucrose and for available Fe treatments 50 μM Fe-(III) EDTA per liter of demineralized water, supplemented with 10 g of Plant agar when plants need to be grown on square Petri dishes. Adjust the pH to 5.5 or pH 7.3 using 1M KOH for normal and alkaline conditions, respectively.

2.2.2 Preparing the WCS417 Inoculum

1. Sterile 10 mM $MgSO_4$ in demineralized water.
2. *King's medium B solution* (KB): 20 g Protease Peptone N°3, 1.5 g of $MgSO_4$, 1.2 g of K_2HPO_4, and 10 g of glycerol per liter of demineralized water, supplemented with 15 g of Difco Agar Granulated.
3. Rifampicin stock: 100 mg · mL^{-1} Rifampicin in DMSO.

2.2.3 Harvesting Plant Material for Metabolomics or Gene Expression

1. Sterile MQ water.
2. Liquid nitrogen.

2.2.4 GUS Staining

1. *GUS buffer solution*: 50 mM sodium phosphate buffer (pH 7.0), 10 mM EDTA (pH 8.0), 0.5 mM $K_3[Fe(CN)_6]$, 0.5 mM $K_4[Fe(CN)_6]$, 0.5 mM X-GlcA and 0.1% (v/v) Triton-X per 50 mL of demineralized water.
2. 20% (v/v) glycerol in demineralized water.

2.2.5 RNA Extraction and cDNA Synthesis

1. Sterile MQ water.
2. 10% SDS stock (pH = 7.2, adjust pH with 1 M HCl).
3. *Cell lysis solution*: 2% SDS, 68 mM sodium citrate, 132 mM citric acid, 1 mM EDTA. The pH of the solution should be between 4 and 4.5, adjust pH with 1 M HCl.
4. *Protein-DNA precipitation solution*: 4 M NaCl, 16 mM sodium citrate, 32 mM citric acid.
5. Isopropanol.
6. DNAse I 1 U · μL^{-1}.
7. 10× Reaction Buffer with $MgCl_2$ for DNase I (Thermo Scientific™).
8. 50 mM EDTA.
9. Oligo dT primer (200 nM).
10. 5× RT buffer (Thermo Scientific™, EP0451).
11. 10 mM dNTPs.
12. RevertAid H minus reverse transcriptase 200 U · μL^{-1} (Thermo Scientific™, P0451).

2.2.6 Quantitative Reverse Transcription-Polymerase Chain Reaction (qRT-PCR)

1. Primers of interest.
2. Sterile MQ water.
3. SYBR® Green Master Mix (Applied Biosystems).

3 Methods

3.1 Seed Sterilization and Sowing

1. Put the desired amount of Arabidopsis seeds into a 1.5-mL Eppendorf tube and vapor-sterilize the seeds by placing them, with the lid of the Eppendorf tube open, in a desiccator together with a glass beaker containing a mix of 100 mL of bleach and 3.2 mL of 37% (v/v) HCl for 3.5 h (*see* **Notes 1** and **2**). This must be done in a fume hood.

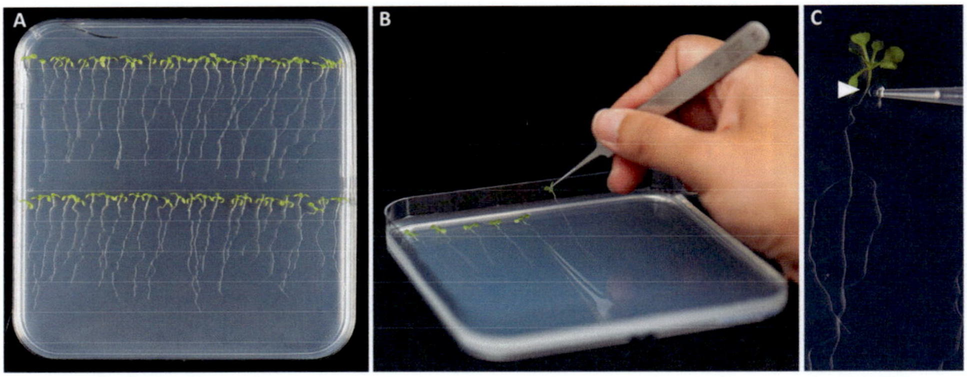

Fig. 1 Square Petri dishes with Arabidopsis plants during different stages of growth and treatment. The sowing pattern of seedlings (**a**), transplantation to a new square Petri dish using sterile tweezers (**b**) and inoculation with *P. simiae* WSC417r at the root-shoot junction (**c**)

2. Close the Eppendorf tubes containing the seeds and move them to a clean flow cabinet. Open the cap of the Eppendorf tube for 30 minutes to allow the excess chlorine gas vapors to disperse out of the tube.
3. *Sowing*:
   (a) For all assays except the coumarin fluorescence quantification assay: sow the seeds on agar-solidified Murashige and Skoog (MS) medium in a square Petri dish (120 × 120 mm) in two rows of 25 seeds (Fig. 1a) and seal the square Petri dish with Parafilm.
   (b) For the coumarin fluorescence quantification assay: sow the seeds on a 96-well plate containing 200 μL of agar-solidified Hoagland medium (with available Fe at pH 5.5 or without available Fe at pH 7.3) in each well, resulting in 1 plant per well. Seal the plate with Parafilm.
4. After sowing, stratify the seeds in darkness at 4 °C for 48 h.
5. Transfer the square Petri dishes and the 96-well plates into the short-day growth chamber and position them vertically and horizontally, respectively.

3.2 Transplanting

1. Take the square Petri dishes containing plants from the growth chamber.
2. *Transplanting*:
   (a) To transfer to square Petri dishes containing agar-solidified medium: open the square Petri dishes containing the plants grown on MS medium and move the plants with sterile tweezers gently to agar-solidified Hoagland medium in square Petri dishes (120 × 120 mm) by lifting the hypocotyls, resulting in 10 plants per square Petri dish (Fig. 1b) (*see* **Note 3**).

Table 1
Overview of plant age for the transplanting and the treatment steps per experiment

Experiment	Transplanting age (days)	Treatment age (days)	Harvest age (days)
GUS assisted localization of gene expression	5	5	10
Preparing plant material for qPCR and coumarin metabolite profiling	10	12	14
Coumarin fluorescence visualization	10	10	17
Coumarin fluorescence quantification	–	–	7

(b) To transfer to 12-well plates containing liquid medium: open the square Petri dishes containing the plants grown on MS medium and move the plants with sterile tweezers gently into a well of a 12-well plate containing 1.5 mL of Hoagland medium per well by lifting the hypocotyls, resulting in 2 plants per well (Fig. 3a).

3. Transfer the square Petri dishes and the 12-well plates to a short-day growth chamber and store them vertically and horizontally, respectively.

See Table 1 for more details.

3.3 Preparing the Bacterial Inoculum and Plant Inoculation

1. Streak bacteria from a frozen glycerol stock on Petri dishes containing agar-solidified KB with 50 μg · mL^{-1} rifampicin using a sterile inoculation loop, and incubate them upside-down, overnight at 28 °C and in darkness.
2. In a sterile flow, add 2–5 mL of sterile 10 mM $MgSO_4$ to the bacterial cultures grown overnight and suspend the bacteria in the solution by scraping them off the agar medium using a sterile spreader.
3. Pipet 50 μL of bacterial suspension to a fresh Petri dish with agar-solidified KB containing 50 μg · mL^{-1} rifampicin and spread them over the medium using a sterile spreader. Store the Petri dish upside-down, overnight at 28 °C and in darkness.
4. After 24 h, add 5 mL of 10 mM $MgSO_4$ to the Petri dish and gently scrape the bacteria from the agar into a suspension using a sterile spreader.
5. Pour the suspension into a 50-mL Falcon tube, then centrifuge at 4500 *g* for 5 min.
6. Pour off the supernatant, then add 25 mL of 10 mM $MgSO_4$ to the tube, and resuspend the pellet using a vortex. Finally, centrifuge the tube at 4500 *g* for 5 min.
7. Repeat **step 6**.

8. Pour off the supernatant and resuspend the pellet in 20 mL of fresh 10 mM $MgSO_4$.
9. Measure the optical density (OD) at $\lambda = 600$ nm of the resulting bacterial suspension (*see* **Note 4**).
10. Then dilute bacterial suspension down to $OD_{600} = 0.1$.
11. Take the square Petri dishes with plants from the growth chamber at your growth stage of interest (Table 1) and open them in a sterile flow cabinet.
12. Using sterile tips, pipet 5–10 μL of bacterial solution or 10 mM $MgSO_4$, as a mock treatment, to the shoot-root junction (Fig. 1c, white arrowhead).
13. Leave the square Petri dishes open to dry but close them immediately as the liquid is evaporated to protect the plants from drying out.
14. Seal the square Petri dishes using Parafilm and store them upright in a short-day growth chamber.

3.4 Harvesting Plant Material for qPCR and Coumarin Metabolite Profiling

1. Sample size for RNA extraction or coumarin metabolomic profiling: for one biological repeat, take 30 roots of plants from 3 independent square Petri dishes (the weight of 30 roots of 14-days-old Arabidopsis is around 50–65 mg) (*see* **Note 5**).
2. Prepare liquid nitrogen, Eppendorf tubes, tweezers (clean with 70% ethanol before use), two square Petri dishes filled with sterile MQ water, and clean tissues.
3. Cut the plants with a razor blade at the shoot-root junction to separate the root from the aerial part of the plant.
4. Carefully collect all roots from three square Petri dishes using tweezers and wash them twice by submerging them in MQ water-filled square plates (*see* **Notes 6** and 7).
5. Gently dry the roots on clean tissues.
6. Transfer the root material into the Eppendorf tubes.
7. Add 3 sterile glass beads into each sterile 1.5 mL Eppendorf tube containing the plant samples.
8. Snap-freeze the samples in liquid nitrogen and store them at −80 °C until RNA isolation or extraction for metabolomic analysis.

3.5 GUS-Assisted Localization of Gene Expression

1. Use 10-day-old plants that have been transplanted after 5 days and treated for 5 days.
2. Prepare a 12-well plate, adding 1 mL of GUS buffer solution to each well.

3. Gently lift the plants off the agar using tweezers and submerge them in sterile MQ. Swivel them around gently to make sure the plants are clean (*see* **Note 7**).
4. Gently submerge the plants in the GUS buffer solution, with a maximum of 5 plants per well. Take care to separate plants of different treatment conditions in separate wells and mark the wells carefully. Make sure no roots and leaves are sticking to the sides of the wells.
5. *Optional*: The 12-well plate can be put in a desiccator and under vacuum for 30 s. This can enhance the GUS staining effectivity as it helps the GUS staining buffer to penetrate the tissue, especially in the leaves. For root tissue of Arabidopsis, it is not necessary, so we often skip this step.
6. Seal the 12-well plate using Parafilm, wrap it in aluminum foil, and store at 37 °C for the required amount of time (*see* **Note 8**).
7. When differences between treatments can be clearly observed, remove the 12-well plate from the stove and carefully remove the GUS buffer solution using a P1000 pipette. A good practice is to tilt the 12-well plate and pipet in a place where no tissue is present.
8. To decolor the plant tissue, add 1 mL of 96% ethanol to each of the wells, seal the 12-well plate using Parafilm, and wrap in aluminum foil. Then place the 12-well plate on a Bio Dancer (New Brunswick) or an equivalent plate shaker. Refresh the ethanol 2 or 3 times (every few hours), the 12-well plate can also be left overnight on the shaker.
9. When the plant tissue has lost all chlorophyll, the plants are ready for imaging. If the plants need to be kept for later analysis, replace the 96% ethanol with 70% ethanol and store them at 4 °C and in darkness.
10. Take the plants out of their wells and submerge them in a container with MQ water and gently untangle them using tweezers.
11. Prepare a microscope slide by adding 100 μL of 20% (v/v) glycerol in MQ to the slide. Place one plant on the slide and take care to spread the root system out, so all roots are clearly visible.
12. Add a cover slip on top by placing its edge at one side of the microscope slide and slowly letting the other edge come down upon the slide. Be careful to avoid formation of air bubbles.
13. Now the slides are ready for microscope imaging (*see* **Note 9**) (Fig. 2).

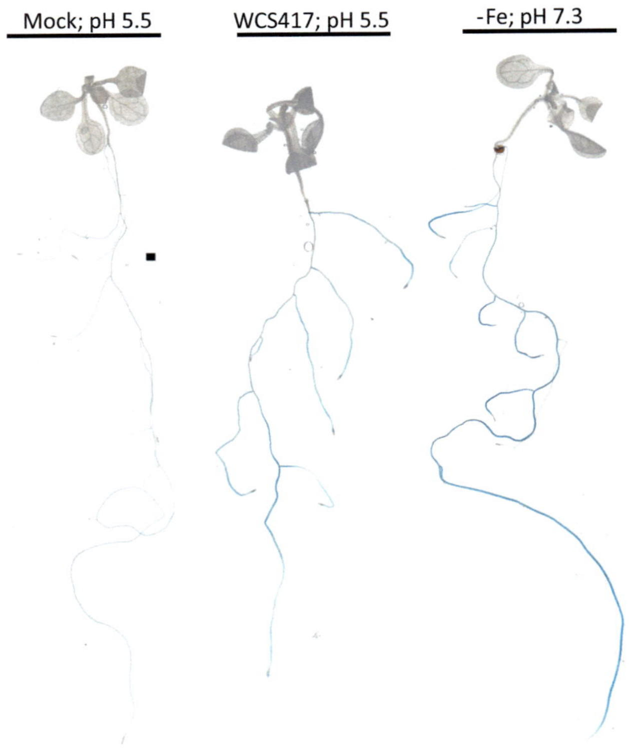

Fig. 2 GUS-stained Arabidopsis *pF6'H1::GUS* plants treated with mock, WCS417 or Fe deficiency conditions. These are composite images of photos that have been stitched using the MosaicJ plugin of ImageJ (*see* **Note 9**)

3.6 Coumarin Fluorescence Visualization

1. Take plants that were transplanted when 10 days old and treated for 7 days from the growth chamber. Remove the lids of the square Petri dishes containing plants of interest and place them on a UV transilluminator (365 nm) in a dark room.
2. Take pictures in only UV light using a photo camera (Fig. 3b, c) (*see* **Note 10**).

3.7 Coumarin Fluorescence Quantification

1. Take the 96-well plates containing 7-day-old plants from the growth chamber.
2. Gently remove the plants from the wells using a tweezer, being careful not to break the roots.
3. Measure the fluorescence of the wells using a microplate reader with a fluorescence detector (excitation at 360 nm; emission at 528 nm).

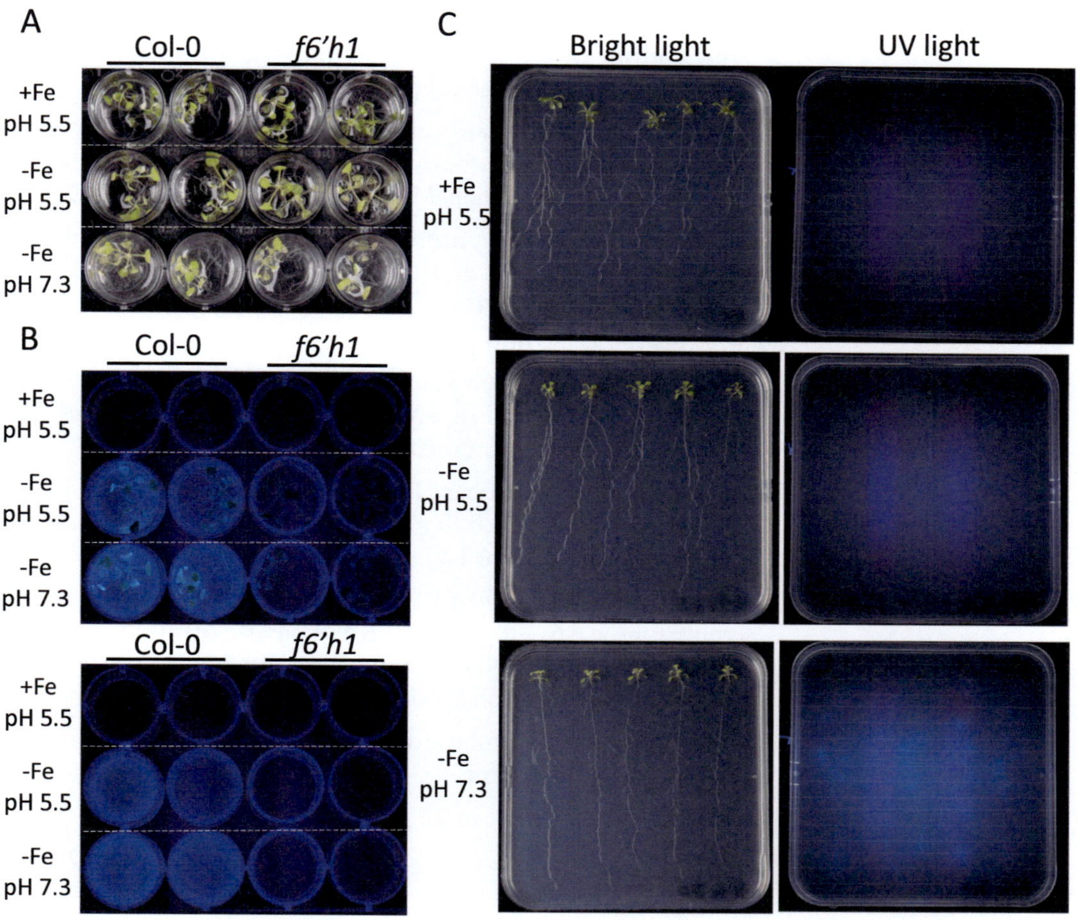

Fig. 3 Visualization of fluorescent coumarins secreted by Arabidopsis in liquid and agar-solidified medium system. (**a**) 10-days-old Col-0 and coumarin biosynthesis deficient mutant *f6′h1* plants growth for 7 days in Hoagland's liquid medium with Fe at pH 5.5 (nonlimited condition), without Fe at pH 5.5 (Fe limited) or without Fe at pH 7.3 (Fe limited plus alkaline stress) (bright light). (**b**) Secretion of fluorescent coumarins of Col-0 and *f6′h1* plants under different Fe-availability treatments (UV light, up: with plants; down: without plants). (**c**) Secretion of fluorescent coumarins of 17-days-old Col-0 on agar-solidified medium system under different Fe-availability treatments (left: bright light; right: UV light)

3.8 RNA Extraction and DNase Treatment

1. Harvest plant materials following the steps shown in Subheading 3.4.
2. The following method is adapted from [23] with slight modifications for use in our system.
3. Pre-cool the rack for TissueLyser II (Qiagen) at −80 °C for at least 5 min.
4. Grind the samples with glass beads using a TissueLyser II (Qiagen) at a frequency of 28 Hz for 30 s (put the samples back into liquid nitrogen immediately after this step, *see* **Note 11**).

5. Take each sample carefully out of the liquid nitrogen and tap the bottom of the tube strongly on a table to avoid ground sample adhering to the lid of the tube.
6. Add 300 μL of cell lysis solution, homogenize rapidly (vortex, tap, and invert the tube); incubate at room temperature for 5 min on rotating Intelli-Mixer™ RM-2 M (*see* **Note 12**).
7. Add 100 μL of protein-DNA precipitation solution (pre-cooled on ice) and homogenize (tap and invert the tube); incubate at 4 °C (on ice) for 10 minutes (*see* **Note 13**).
8. Centrifuge at 16,000 *g* and 4 °C for 15 min; transfer supernatant (300 μL) to a new tube. If some debris/tissue remains in the supernatant, spin down again (16,000 *g*, 5 min) and take the supernatant to a new tube.
9. Add 300 μL of isopropanol and homogenize (invert the tube), then centrifuge at 16,000 *g* and 4 °C for 5 min and pour off the supernatant (*see* **Note 14**).
10. Wash pellet with 300 μL of 70% ethanol (invert the tube); centrifuge at 16,000 *g* for 1 min and pipette off all supernatant, using a pipette (P200) to remove the smallest drops. If necessary, spin down. Wash again as above.
11. Open the lid of the tube, and air dry the pellet for 10 min until it becomes transparent.
12. Resuspend the pellet in 25 μL MQ.
13. Determine the RNA concentration by using NanoDrop™ 2000/2000c.
14. Take 2.2 μg of RNA to a new tube.
15. Add 2 μL 10× DNAse buffer and 1 μL DNAse I (Thermo Scientific™); add RNAse-free MQ to a total volume of 20 μL. Incubate for 30 min at 37 °C.
16. Add 1 μL of EDTA (50 mM); incubate at 65 °C for 10 min to inactivate DNAse.
17. Store the DNAse-treated RNA sample at −80 °C until further use.

3.9 cDNA Synthesis

1. Take 12 μL of DNAse-treated RNA (about 1 μg of RNA) to a new tube of a PCR tube strip (*see* **Note 15**).
2. Add 1 μL of oligo dT primer.
3. Incubate the samples at 65 °C for 5 min.
4. Place the samples on ice immediately.
5. Add 4 μL of 5× buffer, 2 μL of 10 mM dNTPs, and 1 μL RevertAid H minus Reverse Transcriptase, and mix gently (*see* **Note 16**).

6. Incubate the samples at 42 °C for 60 min, then at 70 °C for 15 min.
7. Store the samples at −20 °C until further use.

3.10 Quantitative Reverse-Transcription Polymerase Chain Reaction (qRT-PCR)

1. This protocol is for a 384-well plate loading design with 5 μL total volume of reaction per well.
2. Prepare the sample loading design according to the 384-well plate (with two technical replicates for each sample).
3. Make primer working solution: 25 μL Forward primer (100 μM), 25 μL Reverse primer (100 μM), and 450 μL sterile MQ water (end concentration of primer is 500 nM).
4. Make SYBR® Green Master Mix: for one reaction, use 2.5 μL SYBR® Green and 0.5 μL primer working solution. Prepare the master mix for 10% extra reactions to account for pipetting errors. Divide the master mix into the PCR tube strip. This facilitates the loading of the plate using a multichannel pipette (5–10 μL).
5. Dilute the cDNA 4× in a new PCR tube strip using MQ water (end dilution in the wells is 10×).
6. Before sample loading, spin down all the master mix and cDNA solution and make sure there are no bubbles in the tubes.
7. Take a 384-well plate and place on the bench table (use dark paper underneath the plate for contrast, it helps to load small volumes in each well).
8. Add 3 μL of master mix to the well using the multichannel pipette and low-retention filter tips.
9. Add 2 μL cDNA template to each well using the multichannel pipette and low-retention filter tips.
10. Seal the plate after loading all samples.
11. Spin down using the well-plate centrifuge (1000 *g*, 1 min).
12. Load the plate on a ViiA7 real-time PCR system.
13. qRT-PCR program:
    (a) Initial denaturation: 50 °C for 2 min, 95 °C for 3 min.
    (b) PCR cycle (40 cycles): 95 °C for 15 s, 60 °C for 1 min.
    (c) Melt curve cycle (1 cycle): 95 °C for 15 s, 60 °C for 1 min, 95 °C for 15 s.
14. Calculate transcript levels related to reference gene *At1g13320* using the $2^{-\Delta Ct}$ ($\Delta Ct = (Ct_{\text{target gene}} - Ct_{\text{reference gene}})$ method [24].

4 Notes

1. The optimal duration of vapor sterilization with chlorine gas is between 3 and 4 h. An exposure time above 4 h will be harmful to Arabidopsis seed germination rate.
2. Since the chlorine gas forms immediately after adding the HCl together with the bleach, it is paramount to close the air-tight desiccator immediately after mixing HCL and bleach.
3. When transplanting, make sure the roots of the plants touch the new plate first, after which you will slide the plant upwards to the final resting position of the shoot. This way the roots will be straightened and aligned parallel to each other.
4. Before measuring the OD_{600} we dilute the bacterial suspension 10×. This is because at higher OD values the spectrophotometer may become less accurate.
5. For detection of coumarins using HPLC a minimum of 50 mg of root tissue per sample is advised (although you can go as low as 10 mg). For shoots, a minimum of 100 mg is advised (although you can go as low as 50 mg).
6. Try to collect root samples using tweezers by lifting up root materials instead of clipping, and use new washing plates between treatments.
7. Make sure to especially clean the plants treated with bacteria thoroughly, due to biofilm formation bacteria tend to stick quite strongly to the root.
8. The required amount of time for GUS staining can vary strongly depending on promoter activity and GUS line (15 min to a few hours). Since GUS staining is a cumulative method, all roots will turn blue after sufficient time. Each GUS line should be tested to find the optimal time that allows for detection of differences between treatments in the conditions of interest.
9. Some microscopes have a tile function with which you can make high-resolution pictures of a large area of the root system, but if this equipment is lacking there exists a plugin for ImageJ (available for download at https://imagej.nih.gov/ij/) called MosaicJ (available for download at http://bigwww.epfl.ch/thevenaz/mosaicj/) which allows you to manually stitch images together to approximate the results from tiling [25, 26].
10. When taking pictures, make sure you use a prolonged exposure time. Additionally, as seen in Fig. 3b, c, fluorescent coumarins appear in –Fe conditions in liquid medium at both high and low pH, but on agar plates only at high pH. This could be due to traces of Fe in the plant agar. To account for this, 300 μM of

ferrozine (Sigma) could be added to the medium to make all traces of Fe unavailable to the plant.

11. To inactive RNase, plant materials for RNA extraction need to be stored at −80 °C before extraction and use liquid nitrogen to transport samples before cell lysis solution is added.
12. The pH of cell lysis solution should be between 4 and 4.5; the low pH is important to inactivate RNase.
13. The "cell lysis solution" and "protein-DNA precipitation solution" can be stored at room temperature for up to 1 month.
14. After centrifuging, you will see the pellet at the bottom of tube, use pipette (P1000) to remove supernatant carefully.
15. For faster sample loading later in qRT-PCR, use PCR tube strips (8×) for cDNA synthesis.
16. Prepare master mix containing 5× buffer, 10 mM dNTPs according to the amount of sample before **step 3**, and add transcriptase to master mix during **step 3**.

Acknowledgments

The authors would like to thank Hans van Pelt for the high quality pictures of the experiments and the members of the Plant-Microbe Interactions group who had developed some of the preceding protocols. This work is supported by NWO Gravitation Grant no. 662 024.004.014.

References

1. Riaz N, Guerinot ML (2021) All together now: regulation of the iron deficiency response. J Exp Bot 72(6):2045–2055
2. Grillet L, Schmidt W (2019) Iron acquisition strategies in land plants: not so different after all. New Phytol 224(1):11–18
3. Verbon EH et al (2017) Iron and immunity. Annu Rev Phytopathol 55:355–375
4. Aznar A, Dellagi A (2015) New insights into the role of siderophores as triggers of plant immunity: what can we learn from animals? J Exp Bot 66(11):3001–3010
5. Dellagi A et al (2005) Siderophore-mediated upregulation of Arabidopsis ferritin expression in response to Erwinia chrysanthemi infection. Plant J 43(2):262–272
6. Trapet PL et al (2021) Mechanisms underlying iron deficiency-induced resistance against pathogens with different lifestyles. J Exp Bot 72(6):2231–2241
7. Kloepper JW et al (1980) Pseudomonas siderophores – a mechanism explaining disease-suppressive soils. Curr Microbiol 4(5): 317–320
8. Pieterse CMJ et al (2021) Pseudomonas simiae WCS417: star track of a model beneficial rhizobacterium. Plant Soil 461:245–263
9. Pieterse CMJ et al (2014) Induced systemic resistance by beneficial microbes. Annu Rev Phytopathol 52:347–375
10. Stassen MJJ et al (2021) Coumarin communication along the microbiome–root–shoot axis. Trends Plant Sci 26(2):169–183
11. Verbon EH et al (2019) Rhizobacteria-mediated activation of the Fe deficiency response in Arabidopsis roots: impact on Fe status and signaling. Front Plant Sci 10:909
12. Zamioudis C et al (2015) Rhizobacterial volatiles and photosynthesis-related signals coordinate MYB72 expression in Arabidopsis roots during onset of induced systemic resistance

and iron-deficiency responses. Plant J 84(2): 309–322

13. Van der Ent S et al (2008) MYB72 is required in early signaling steps of rhizobacteria-induced systemic resistance in Arabidopsis. Plant Physiol 146(3):1293–1304
14. Zamioudis C, Hanson J, Pieterse CMJ (2014) β-Glucosidase BGLU42 is a MYB72-dependent key regulator of rhizobacteria-induced systemic resistance and modulates iron deficiency responses in Arabidopsis roots. New Phytol 204(2):368–379
15. Rosenkranz T et al (2021) Root exudation of coumarins from soil-grown Arabidopsis thaliana in response to iron deficiency. Rhizosphere 17:100296
16. Schmid NB et al (2014) Feruloyl-CoA 6′--Hydroxylase1-dependent coumarins mediate iron acquisition from alkaline substrates in Arabidopsis. Plant Physiol 164(1):160–172
17. Stringlis IA et al (2018) MYB72-dependent coumarin exudation shapes root microbiome assembly to promote plant health. Proc Natl Acad Sci U S A 115(22):E5213–E5222
18. Tsai HH et al (2018) Scopoletin 8-hydroxylase-mediated fraxetin production is crucial for iron mobilization. Plant Physiol 177(1): 194–207
19. Harbort CJ et al (2020) Root-secreted coumarins and the microbiota interact to improve iron nutrition in Arabidopsis. Cell Host Microbe 28(6):825–837
20. Voges M et al (2019) Plant-derived coumarins shape the composition of an Arabidopsis synthetic root microbiome. Proc Natl Acad Sci U S A 116(25):12558–12565
21. Murashige T, Skoog F (1962) A revised medium for rapid growth and bio assays with tobacco tissue cultures. Physiol Plant 15(3): 473–497
22. Hoagland DR, Arnon DI (1938) The water-culture method for growing plants without soil. Calif Agric Exp Stn Bull 347:36–39
23. Onate-Sanchez L, Vicente-Carbajosa J (2008) DNA-free RNA isolation protocols for Arabidopsis thaliana, including seeds and siliques. BMC Res Notes 1:93
24. Czechowski T et al (2005) Genome-wide identification and testing of superior reference genes for transcript normalization in Arabidopsis. Plant Physiol 139(1):5–17
25. Schneider CA, Rasband WS, Eliceiri KW (2012) NIH image to ImageJ: 25 years of image analysis. Nat Methods 9(7):671–675
26. Thevenaz P, Unser M (2007) User-friendly semiautomated assembly of accurate image mosaics in microscopy. Microsc Res Tech 70(2):135–146

Chapter 6

Imaging and Quantifying the Endocytosis of IRON-REGULATED TRANSPORTER1 from Arabidopsis

Julien Spielmann, Julie Neveu, and Grégory Vert

Abstract

Iron plays an essential role in plant metabolism and the regulation of its transport is essential for the plant. In *Arabidopsis thaliana*, iron uptake in root epidermal cells is mediated by the IRT1 (IRON-REGULATED TRANSPORTER 1) broad-spectrum transporter. The regulation of the IRT1 protein is controlled by sophisticated mechanisms that allow it to fine-tune the amount of transporter found at the plasma membrane and to modulate the uptake of iron and divalent metals transported by IRT1. IRT1 shows low selectivity and transports different metals such as manganese, zinc, cobalt, and cadmium. An excess of these non-iron metal substrates of IRT1 is toxic for the plant. The ability of plants to adapt to non-iron metal stress is based on the sensing of their excess, leading to the internalization and degradation of IRT1. IRT1 acts as a bifunctional transporter/receptor directly sensing metal non-iron excess and then undergoes a series of post-translational modifications of the protein culminating in its endocytosis and vacuolar degradation. To monitor the intracellular dynamics of IRT1, we describe in this chapter a live cell imaging approach to follow and quantify IRT1-mCitrine trafficking from the plasma membrane to the vacuole.

Key words Iron, Endocytosis, Imaging, Arabidopsis

1 Introduction

To sustain proper growth and development, plants must acquire essential nutrients from the soil [1]. Iron is a crucial co-factor for many biological processes including photosynthesis, cellular respiration, or DNA synthesis, due to its capacity to easily change oxidative state. Although iron is the fourth most abundant ion in the earth's crust, it is poorly available to plants, and iron deficiency stands out as one of the major constraints in modern agriculture [2–4]. This is especially true in alkaline or aerobic conditions where iron tends to precipitate to form insoluble ferric complexes [2]. To acquire iron from the soil, plants evolved and developed two different strategies. Strategy I, which occurs in all plant species except grasses, consists of a sophisticated multistep iron uptake strategy relying on local rhizosphere acidification, iron reduction, and

Jeeyon Jeong (ed.), *Plant Iron Homeostasis: Methods and Protocols*, Methods in Molecular Biology, vol. 2665, https://doi.org/10.1007/978-1-0716-3183-6_6,

ferrous iron transport into the root cells. For the sake of this chapter dedicated to the endocytosis of the major *Arabidopsis thaliana* root iron transporter, we will only describe strategy I.

In the model plant *A. thaliana*, the three steps of iron uptake are mediated by a high-affinity iron complex located at the plasma membrane. This complex is composed of three members: (i) the proton pump H^+-ATPase2 (AHA2), which releases proton in the rhizosphere to increase iron solubility, (ii) the FERRIC REDUCTION OXYDASE2 (FRO2) that reduces ferric iron into ferrous iron at the root surface, and (iii) the IRON-REGULATED TRANSPORTER1 (IRT1) that transports ferrous iron into roots epidermal cells [5–7]. Perturbation of one of these processes strongly impacts plant growth and yields severe chlorosis, demonstrating their essential function in iron uptake.

Upon iron-limited conditions, *IRT1* transcription is rapidly and strongly induced in the root epidermis leading to an increased accumulation of IRT1 protein and enhanced iron uptake [8]. However, due to its low metal selectivity, *IRT1* induction also boosts the absorption of non-iron metals such as zinc, manganese, cobalt, or cadmium (hereafter called non-iron metals) [8–10]. To limit the accumulation of highly reactive and potentially toxic non-iron metals upon low iron, plants evolved a complex regulatory mechanism by locally adjusting the levels of IRT1 at the cell surface [11]. IRT1 acts as a bifunctional transporter and receptor, capable of direct binding and sensing non-iron metals through a histidine-rich motif located in a cytoplasmic regulatory loop [11, 12]. In the absence of non-iron metals, IRT1 is localized mainly at the plasma membrane where it actively transports traces of iron. An Increase in non-iron metals gradually shifts IRT1 subcellular localization from the plasma membrane to the early endosome and increases IRT1 degradation in the vacuole. The direct interaction between non-iron metals and the histidine-rich motif of IRT1 triggers the recruitment of the CIPK23 kinase (CBL-INTERACTING PROTEIN KINASE 23), which phosphorylates serine and threonine residues in the IRT1 regulatory loop [11]. Phosphorylation of IRT1 has two distinct effects. First, it mediates the dissociation of the AHA2/FRO1/IRT1 high-affinity iron uptake complex [7]. Secondly, it creates a docking site in IRT1 for the IDF1 (IRON DEGRADATION FACTOR 1) RING E3 ubiquitin ligase, leading to the decoration of IRT1 with K63-linked polyubiquitin chains attached to lysine residues K159 and K174 [11, 13]. Polyubiquitination of IRT1 drives its endosomal sorting and vacuolar targeting where IRT1 is degraded [11]. Such local and rapid adjustment of IRT1 levels allows plants to optimize iron uptake while reducing the accumulation of non-iron metals [11, 12, 14].

In this chapter, we provide detailed protocols to investigate the endocytic dynamics of IRT1 in response to non-iron metal exposure.

2 Materials

2.1 Plant Materials

Arabidopsis thaliana expressing a functional IRT1-mCitrine fusion. To facilitate the imaging of IRT1 endocytosis and to specifically focus on post-translational regulatory events controlling IRT1 endocytosis, a transgenic line expressing IRT1-mCitrine (IRT1-mCit) under the control of the PIN2 promoter [15]. For the vacuolar targeting experiment, plants co-expressing IRT1-mCit under constitutive 35S promoter (35S::IRT1-mCit) [11] and the tonoplast marker mRFP- SYP22 [16] were used.

2.2 Medium Preparation

- Murashige and Skoog basal salt macronutrient solution.
- Murashige and Skoog basal salt micronutrient solution without iron and non-iron metals (Table 1).
- Non-iron metals solution (Table 2).
- MES hydrate.
- Sucrose.
- Agar Type M.
- Mili-Q water.
- 1 M Potassium hydroxide 1 M.
- 12 cm square Petri dishes.
- 12 wells plates.

Table 1
Murashige and Skoog basal salt micronutrient solution without iron and non-iron metals in 2000X final concentration

Formula	Name	MW (g.mol^{-1})	Concentration in stock solution (2000X)
H3BO3	Boric acid	61.83	100 mM
CuSO4, 5H2O	Cupper Sulfate	249.7	100 μM
Na2MoO4, 2H2O	Sodium Molybdate	241.95	1 mM
KI	Potassium iodide	166	5 mM

Table 2
Non-iron metals solutions in 2000X final concentration

Formula	Name	MW (g.mol^{-1})	Concentration in stock solution (2000X)
MnSO4, H2O	Manganese Sulfate	169.02	100 mM
ZnSO4, 7H2O	Zinc Sulfate	287.53	30 mM
CoCl2, 6H2O	Cobalt Chloride(II)	237.93	100 μM

2.3 Additional Materials

- 98% EtOH.
- Seed sterilization solution: 70% EtOH, 0.05% Triton-X100.
- Microporous tape.
- Microscope slide and cover glass.

2.4 Lab Equipment

- Vertical holder for square Petri dishes.
- Sterile hood.
- Plant growth chamber.
- Fridge or cold room.
- pH meter.
- Weighting machine.
- Autoclave.
- Pipette.
- Forceps.
- Confocal scanning microscope.

3 Methods

In this chapter, we provide a detailed analysis of IRT1 subcellular localization in response to non-iron metals or dark treatment. Non-iron metal excess triggers IRT1 endocytosis and vacuolar degradation. Dark treatment limits the lytic activity of the vacuole and therefore allows for better visualization of the targeting of IRT1 to the vacuole (Fig. 1) [11].

3.1 Sample Preparation

This section describes all steps necessary to obtain *A. thaliana* roots displaying IRT1-mCit accumulation at the plasma membrane.

1. Prepare liquid and solid media [1/2 MS lacking iron and non-iron metals (zinc, cobalt, manganese), hereafter called −/− media] in a 500 mL bottle (*see* **Note 1**):
   (a) For a 500 mL bottle, add:
      - 5 g of Sucrose
      - 0.25 g of MES hydrate
      - 25 mL of Murashige and Skoog basal salt macronutrient solution
      - 250 μL of Murashige and Skoog basal salt micronutrient solution without iron and non-iron metals
      - 450 mL of MilliQ water
   (b) Using KOH 1 M, adjust pH to 5.7.

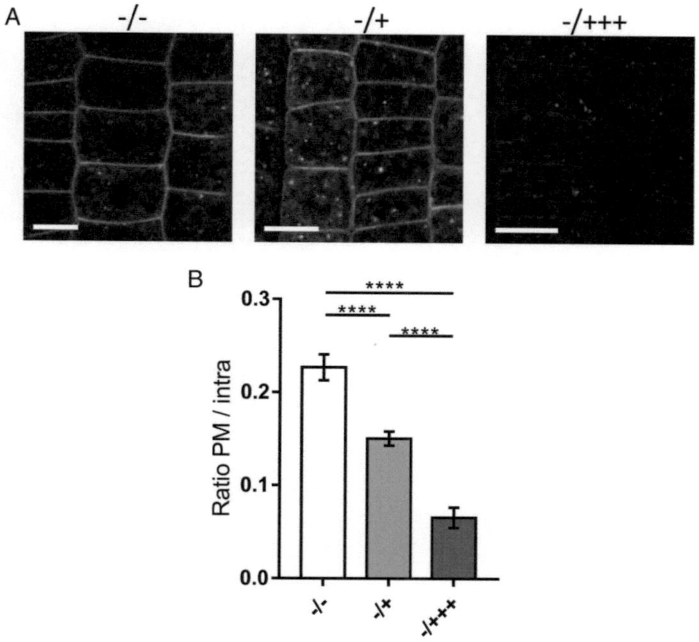

Fig. 1 Non-iron metals regulate IRT1 endocytosis. (**a**) Confocal microscopy analyses of root epidermal cells from *A. thaliana* plants expressing IRT1-mCit. Plants were grown in −/− medium, as explained in this chapter, and liquid-treated for 24 h with different concentrations of non-iron metals as described (−/−; −/+; −/+++). Scale bars, 10 μm. (**b**) Quantification of the ratio of plasma membrane to intracellular signals. Error bars represent SEM (N = 14). Asteriks indicate significant differences between conditions (one-way ANOVA, Tukey post-test, **** $p < 0.0001$)

(c) *Only for solid medium*: Add 5 g of Agar Type M.

(d) Fill up to 500 mL with MilliQ water.

(e) Sterilize media by Autoclaving.

2. Sterilize at least 50 seeds of the plant expressing IRT1-mCit (*see* **Note 2**). For this purpose, place IRT1-mCit seeds in a 1.5 mL microfuge tube. Add 500 μL of sterilization solution and shake for 10 min on a rotation wheel. Remove the tube from the wheel and wait for the seeds to sediment. In a sterile hood, remove the sterilization solution and wash the seeds twice with 98% EtOH. Then, remove all traces of EtOH and leave the tube open until all seeds are dry (around 30 min).

3. In parallel, prepare one plate of −/− medium. In a sterile hood, pour around 50 mL of melted solid −/− medium into a 12 cm square Petri dish. When the medium is solid, saw around 50 surface-sterilized seeds on the plate. Close the plate and wrap it with microporous tape.

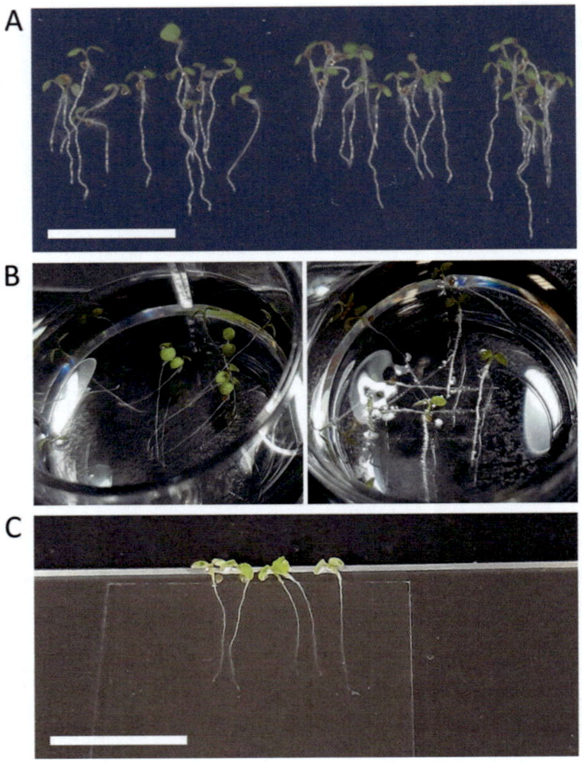

Fig. 2 Different stages of this protocol. (**a**) Photo showing 5-day-old seedlings growing on vertical plate. (**b**) Photo showing seedlings transferred into a 12-well plate for liquid treatment. On the left, seedlings are submerged into the liquid medium (optimal). On the right, seedlings are floating on the medium, surface tension is visible (not optimal). (**c**) Photo showing seedlings transferred on a microscopy slide for analysis. Scale bars, 1 cm

4. For stratification, keep the plate in the dark for at least 2 days at 4 °C.
5. Using a Petri dish holder, place the plate vertically in a climate-controlled growth chamber and leave the seed to grow for 5 days (Fig. 2a).

3.2 Non-iron Metals Treatment and IRT1 Subcellular Localization

This section explains how to treat IRT1-mCit plants with non-iron metals and how to quantify the observed effects.

1. In a 12-well plate, put 2 mL of liquid −/− medium in the appropriate number of wells. Delicately transfer 5–10 seedlings from the plate to the same well using forceps without damaging the roots (Fig. 2b). Add the appropriate amount of non-iron metals to obtain the desired non-iron metal condition. In our previous studies, 3 conditions were used, lack of non-iron metals (−/−), physiological concentrations of non-iron metals (−/+), and 11-fold excess of non-iron metals (−/+++)

[11]. These conditions respectively correspond to the addition of 0, 1, and 11 μl of each non-iron metals stock solution (Subheading 2.2).

2. To obtain a homogenous treatment, mix by gently pipetting until roots are totally submerged (*See* **Note 3**) (Fig. 2b).
3. Put back the 12 wells plate in the climate-controlled growth chamber and incubate it for 2–24 h. Treatments at 2 h and 24 h are each used to analyze fast and slow responses to non-iron metals, respectively.
4. Using forceps, delicately lay 3–5 seedlings on a microscope slide, mount them with around 100 μL of the treatment medium, and put a cover slip (Fig. 2c). Remove the excess liquid using an absorbing paper.
5. Using a confocal scanning microscope, immediately analyze mCitrine signal intensity and localization (*see* **Notes 4** and **5**). Image cells from the transition zone where the lateral root cap ends, and select cells with the same shape, size, and depth along the root to ensure the reproducible and faithful comparison between replicates, treatments, or genotypes (Fig. 3a) (*See* **Note 6**).
6. For eventual quantification, use the confocal tool Z-stacks, and take one image of each micrometer in the Z axis. Be sure to encompass the whole cell volume to detect all endosomes.
7. Using LAS X or ImageJ software and all Z-stacks encompassing the whole cell volume, proceed with a maximal projection.
8. Determine the mean fluorescence of the whole cell and the intracellular content. To do so, use the "polygon selection" tool from the ImageJ software to draw two polygons, one

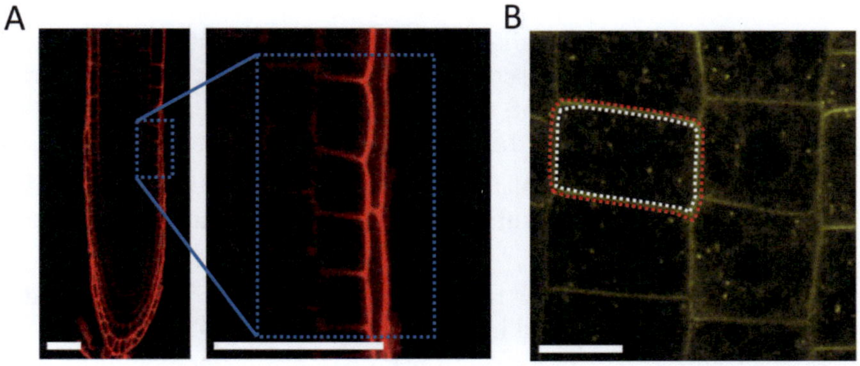

Fig. 3 Cells chosen for imaging and analysis. (**a**) Confocal microscopy analyses of longitudinal section of an *A. thaliana* root stained with propidium iodide. Blue rectangle show the optimal zone to analyze IRT1 subcellular localization. Scale bar 50 μm. (**b**) Confocal microscopy analyses of root epidermal cells from *A. thaliana* expressing IRT1-mCit. White and red polygons show intracellular area and whole cells aera respectively used to quantify the ratio of plasma membrane over intracellular signals. Scale bars, 10 μm

closely surrounding the plasma membrane of the selected cell (Fig. 3b, red polygon) and the second close to the inside border of the plasma membrane (Fig. 3b, white polygon).

9. Finally, determine the ratio of plasma membrane over intracellular signal using the following equation:

$$\frac{\text{Whole cell signal} - \text{Intracellular signal}}{\text{Intracellular signal}}$$

10. To obtain relevant quantification results, we suggest using at least 5 cells of 3 different roots during 3 independent experiments.

3.3 Dark Treatment and IRT1 Subcellular Localization

This last protocol section explains how to perform dark treatment and how it can be used to highlight the impact of a candidate protein or a specific condition on IRT1 trafficking. For dark treatment, plants co-expressing IRT1-mCit and a tonoplast marker such as mRFP-SYP22 can be used to facilitate the visualization of the vacuole.

Two different 12-well plates must be used, one for light and the second for the dark treatment.

1. Perform a duplicate liquid treatment using the light-dedicated plate (*see* **Note** 7), as explained in Subheading 3.2, **step 1** (Fig. 2). The dark-dedicated plate will be used later in **step 4**.
2. Even if plants are not subjected to non-iron metal treatment, make sure plants are submerged by gently pipetting up and down (*See* **Note 3**) (Fig. 2).
3. Place back the light 12-well plate in the climate-controlled growth chamber and incubate it for 20–21 h.
4. 3–4 h before confocal observation, transfer around 5–10 seedlings from one of the light-dedicated 12-well plate duplicates to the dark-dedicated plate and incubate in complete darkness until observation (*see* **Note 8**). At the same time, a metal treatment can be performed (as described previously) to increase the amount of IRT1-mCit targeting the vacuole.
5. Mount plants on the microscope slide as described previously, taking caution to use minimum lighting during the procedure (Fig. 2).
6. Confocal microscopy analyses can be performed exactly as described in Subheading 3.2, **step 5** (Fig. 4).

   Eventually, the ratio of plasma membrane over intracellular signal, or the more meaningful intracellular signal alone (Fig. 3b, only white polygon) can be quantified, as described in Subheadings 3.2, **steps 6–10**.

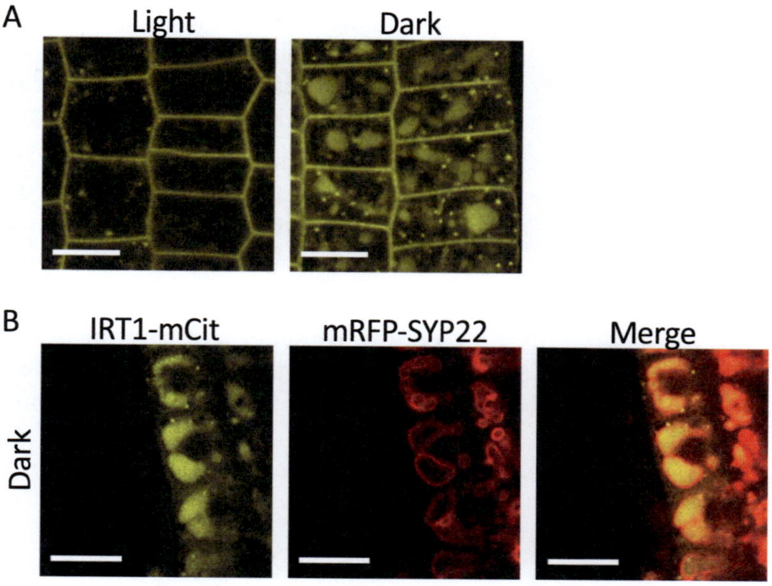

Fig. 4 Impact of dark treatment on IRT1 imaging. (**a**) Confocal microscopy analyses of root epidermal cells from *A. thaliana* plants expressing IRT1-mCit. Plants were grown and liquid-treated in −/− medium, as explained in this chapter. Then, plants were dark incubated (Dark) or not (Light) for 3 h. (**b**) Confocal microscopy analyses of dark-treated IRT1-mCit plants crossed to the tonoplast-localized mRFP-SYP22 marker (Ebine et al., 2008) to reveal fluorescence in the vacuole. Plants were grown −/− medium and liquid reated in −/+ ++ medium in dark for 4 h, as explained in this chapter. Scale bars, 10 μm

4 Notes

1. Solid and liquid mediums can be prepared in excess and use for different experiments.
2. Seeds can be sterilized in larger amounts and conserved for at least 6 months without alteration of germination rate.
3. Once seedlings have been transferred to the liquid medium, they tend to float. Breaking the surface tension by pipetting is essential to obtain homogenous treatment (Fig. 2b).
4. In low light condition, IRT1-mCitine yields a strong vacuolar fluorescence accumulation due to lower lytic activity of the vacuole. The time between taking plants out of the climate-controlled growth chamber and confocal microscopy imaging should be minimized to enhance result quality.
5. Using a 63 folds magnification objective is optimal for detailed observation of endosomal compartments.

6. For eventual quantification of the plasma membrane over intracellular signal ratio, using cells of the same shape is essential. Due to the calculation method, a rectangle-shaped cell has a higher ratio than a square-shaped cell.
7. −/− liquid treatment is the most appropriate to be combined with dark treatment. In presence of non-iron metals, a part of IRT1 is already transported to the vacuole and degraded. Consequently, dark treatment is less efficient.
8. Full darkness is essential to completely abolish vacuolar degradation. We suggest placing the plates wrapped up in aluminum foil in two different containers (e.g., a cardboard box in a cupboard, or a small carton box into a bigger one).

Acknowledgments

We would like to acknowledge the Fédération de Recherche Agrobiosciences Interactions et Biodiversité of Toulouse (FRAIB) for the access confocal microscopes. This work was supported by research grants from the French National Research Agency (ANR-21-CE20-0046-01 and to G.V.) and the French Laboratory of Excellence (project "TULIP" grant nos. ANR-10-LABX-41 and ANR-11-IDEX- 0002-02 to G.V.).

References

1. Marschner P (1995) Marschner's mineral nutrition of higher plants. Elsevier/Academic Press
2. Briat J-F, Dubos C, Gaymard F (2015) Iron nutrition, biomass production, and plant product quality. Trends Plant Sci 20:33–40. https://doi.org/10.1016/j.tplants.2014.07.005
3. Cox PA (1989) The elements. Their origin, abundance, and distribution. ui.adsabs.harvard.edu
4. Wedepohl KH (1995) The composition of the continental crust. Geochimica et cosmochimica Acta
5. Guerinot ML (2000) The ZIP family of metal transporters. Biochim Biophys Acta 1465:190–198
6. Jeong J, Merkovich A, Clyne M, Connolly EL (2017) Directing iron transport in dicots: regulation of iron acquisition and translocation. Curr Opin Plant Biol 39:106–113. https://doi.org/10.1016/j.pbi.2017.06.014
7. Martín-Barranco A, Spielmann J, Dubeaux G et al (2020) Dynamic control of the high-affinity iron uptake complex in root epidermal cells. Plant Physiol 184:1236–1250. https://doi.org/10.1104/pp.20.00234
8. Vert G, Grotz N, Dédaldéchamp F et al (2002) IRT1, an Arabidopsis transporter essential for iron uptake from the soil and for plant growth. Plant Cell 14:1223–1233. https://doi.org/10.1105/tpc.001388
9. Rogers EE, Eide DJ, Guerinot ML (2000) Altered selectivity in an Arabidopsis metal transporter. Proc Natl Acad Sci U S A 97:12356–12360. https://doi.org/10.1073/pnas.210214197
10. Vert G, Briat JF, Curie C (2001) Arabidopsis IRT2 gene encodes a root-periphery iron transporter. Plant J 26:181–189. https://doi.org/10.1046/j.1365-313x.2001.01018.x
11. Dubeaux G, Neveu J, Zelazny E, Vert G (2018) Metal sensing by the IRT1 transporter-receptor orchestrates its own degradation and plant metal nutrition. Mol Cell 69:953–964.e5. https://doi.org/10.1016/j.molcel.2018.02.009
12. Cointry V, Vert G (2019) The bifunctional transporter-receptor IRT1 at the heart of metal sensing and signalling. New Phytol 223:

1173–1178. https://doi.org/10.1111/nph.15826

13. Barberon M, Zelazny E, Robert S et al (2011) Monoubiquitin-dependent endocytosis of the iron-regulated transporter 1 (IRT1) transporter controls iron uptake in plants. Proc Natl Acad Sci U S A 108:E450–E458. https://doi.org/10.1073/pnas.1100659108
14. Spielmann J, Vert G (2020) The many facets of protein ubiquitination and degradation in plant root iron deficiency responses. J Exp Bot. https://doi.org/10.1093/jxb/eraa441
15. Spielmann J, Cointry V, Devime F, Ravanel S Neveu J, Vert G (2022) Differential metal sensing and metal-dependent degradation of the broad spectrum root metal transporter IRT1. The Plant Journal 112:1252–1265. https://doi.org/10.1111/tpj.16010
16. Ebine K, Okatani Y, Uemura T et al (2008) A SNARE complex unique to seed plants is required for protein storage vacuole biogenesis and seed development of Arabidopsis thaliana. Plant Cell 20:3006–3021. https://doi.org/10.1105/tpc.107.057711

Chapter 7

Label-Free Quantitative Proteomics in Plant

Ruonan Wang, Peijun Zhou, Yilin Pan, Lu Zheng, Xiaoying Dong, Renfang Shen, and Ping Lan

Abstract

Label-free quantitation (LFQ) proteomics, mainly based on the extraction of the peptide (precursor) intensity at the MS^1 (mass spectrum 1) level, enables to quantify the relative amount of the proteins among samples. In an LFQ proteomics study, all samples are scanned individually on an advanced mass spectrometer and the chromatographic features of each run are extracted to generate consensus patterns among various runs in the experiment. Here, we describe the LFQ proteomics experimental protocol adapted for plant research, such as plant iron homeostasis.

Key words Label-free quantitation, Proteomics, Mass spectra, Plant

1 Introduction

Quantitative proteomics can significantly contribute to uncover the possible relationships between changes in protein abundance and the responses to various environmental stresses [1], particularly in the post-genome era. In principle, quantitation proteomics can be divided into two categories, with one being label-dependent such as SIL (Stable Isotope Labeling) [2], TMT (Tandem Mass Tag) [3], iTRAQ (Isobaric Tag for Relative Absolute Quantitation) [4] labeling, and the other being label-free quantitation (LFQ) [5]. Although label-based quantitative proteomics could provide qualitative information about the target proteome, it is expensive and laborious, and has other limitations such as low reproducibility among runs and less efficacy on low-abundance proteins. LFQ applies two quantification strategies either by spectral counting or by precursor intensity measurement to reflect the protein abundance (Fig. 1). LFQ proteomics can determine the relative amount of proteins in two or more biological samples, being a high throughput technique. For an LFQ proteomics project, each sample will be individually scanned by mass spectrometry using the

Jeeyon Jeong (ed.), *Plant Iron Homeostasis: Methods and Protocols*, Methods in Molecular Biology, vol. 2665,
https://doi.org/10.1007/978-1-0716-3183-6_7,

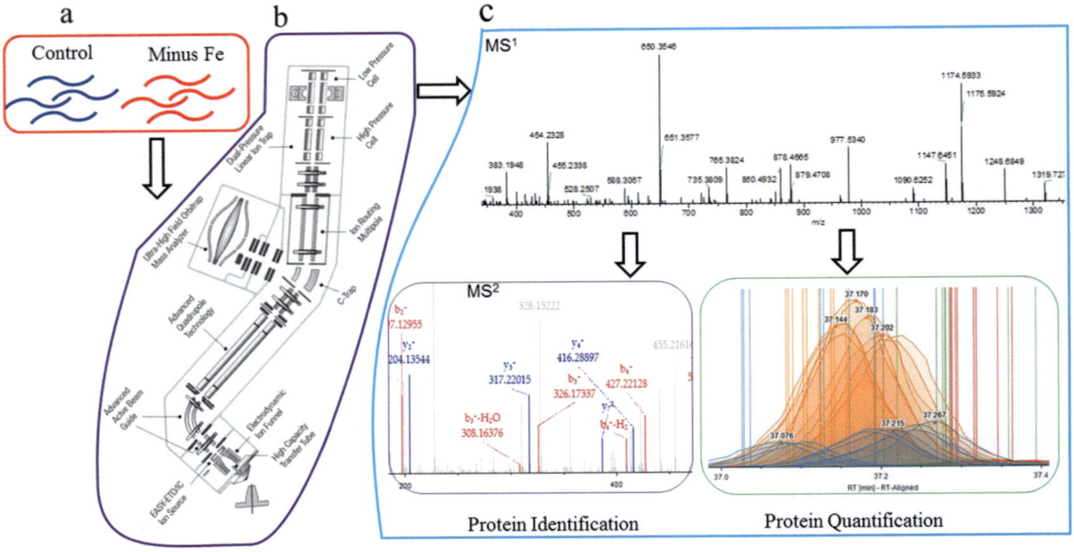

Fig. 1 The workflow of Label-free quantitation (LFQ) proteomics. (**a**) Sample preparation including fresh sample collecting, protein sample extraction and digestion, as well as peptide mixture desalting. (**b**) Nano-LC-MS/MS operation in a mass spectrometer (Fig. 1 indicates the Orbitrap Fusion Lumos mass spectrometer). (**c**) Protein identification and quantification. Protein identification is completed by MS^2 searching against a protein database, while protein quantification mainly depends on the ion intensity of the precursor at MS^1 by aligning the retention time

same protocol. The proteins from each sample are subsequently identified, and the protein abundance from each sample is assessed either by the number of MS/MS spectra identifying the peptide of the protein or by the intensity of the corresponding MS spectrum features of the protein. Spectral counting is easy to analyze after scanning, but the resolution is low because more time is required to run an additional MS/MS scan. Precursor intensity-based LFQ proteomics applies advanced pattern comparison software to align ion intensities of the same peptide on MS^1 and MS^2 levels in different samples. Currently, this method has been becoming the mainstream of proteomics research when LFQ proteomics is adopted. Overall, LFQ proteomics is cost-effective and possesses high-proteome coverage, without sample number limitations, and does not require laborious labeling workflows. Here, we present the LFQ experimental protocol optimized for plant research, which has been validated to run very well in recent studies [6–11].

2 Materials

Ultrapure deionized water with a sensitivity of 18 MΩ.cm at 25 °C is used to prepare solutions. All chemicals are analytical grade and all solutions are stored at room temperature unless indicated otherwise.

2.1 Protein Sample Preparation and Protein Concentration Determination

1. TCA-Acetone solution: 10% trichloroacetic acid (TCA) in acetone with or without 0.07% β-mercaptoethanol, store at −20 °C (*see* **Note 1**).
2. 100% acetone: containing 0.07% β-mercaptoethanol and 1 mM phenylmethanesulfonyl fluoride (PMSF), store at −20 °C (*see* **Note 1**).
3. SDT buffer: 2% sodium dodecyl sulfate (SDS) in 0.1 M Tris containing 0.1 M DL-dithiothreitol (DTT), pH is adjusted to 7.6 with HCl (*see* **Note 2**).
4. Tryptophan solution: 0.5 μg/μL tryptophan in SDT buffer, store at −20 °C.
5. Dilution buffer: 8 M urea in 20 mM Tris-HCl, pH 7.6, store at room temperature (*see* **Note 3**).

2.2 Protein Digestion

1. UA buffer: 8 M urea in 0.1 M Tris, adjust the pH to 8.5, store at room temperature (*see* **Note 3**).
2. Alkylation solution: 50 mM idoacetamide (IAA) in UA buffer, store at room temperature (*see* **Note 4**).
3. NH_4HCO_3 solution: 50 mM ammonium bicarbonate (NH_4HCO_3) in ultrapure deionized water, store at room temperature.
4. Trypsin solution: 0.25 μg/μL trypsin in 10 mM acetic acid, store at −20 °C.

2.3 Peptide Desalt

1. 100% methanol
2. Condition and elution buffer: 80% acetonitrile in ultrapure deionized water, containing 0.2% trifluoroacetic acid (TFA)
3. Equilibration buffer: 0.2% TFA in ultrapure deionized water

2.4 MS/MS Analysis

1. Dionex UltiMate™ 3000 RSLCnano System (Thermo Scientific)
2. Orbitrap Fusion Lumos mass spectrometer (Thermo Scientific)
3. Acclaim PepMap™ 100 column (100 μm × 2 cm, Thermo Scientific)
4. Solvent A: 0.1% formic acid in ultrapure deionized water (*see* **Note 5**)
5. Solvent B: 0.1% formic acid in acetonitrile (*see* **Note 5**)

2.5 Protein Identification and Quantification

1. The Proteome Discoverer software (version 2.3, Thermo Scientific) is used to identify and quantify differentially accumulated proteins (DAPs) simultaneously.

2. Protein database (Arabidopsis, Araport11_genes.201606.pep.fasta; Wheat, Triticum_aestivum IWGSC pep all.fasta; Rice, Oryza_Sativa.IRGSP-1.0.pep.all.fasta).

3 Methods

3.1 Protein Sample Preparation

For plant protein extraction, we suggest to use TCA-Acetone method described by Lan et al [12, 13], at least three biological repeats for each treatment are required.

1. Plant samples are collected in liquid nitrogen immediately and stored at −80 °C.
2. Both sterilized mortars and pestles are precooled in liquid nitrogen.
3. Grind the sample (1 g) into a fine powder in liquid nitrogen (*see* **Note 6**).
4. Transfer the powder into a 50 ml of precooled centrifuge polypropylene bottle.
5. Add 10 × volume of −20 °C stored TCA-Acetone solution and vortex to suspend the powder (*see* **Note 7**).
6. Put the bottle at −20 °C for 2 h.
7. Place the bottles in a high-speed centrifuge rotor (Beckman coulter, Cat. No.JA25-50) and centrifuge the samples at 35,000× *g* at 4 °C for 30 min.
8. Discard the supernatant and resuspend the pellet in 10× volume of 100% acetone.
9. Put the bottle at −20 °C for 1 h and centrifuge at 35,000× *g* at 4 °C for 30 min.
10. Repeat steps 8 and 9 two times (*see* **Note 8**).
11. Dry the pellet under vacuum and transfer the dried pellet to new Eppendorf tube either to next step or to store at −80 °C (*see* **Note 9**).
12. Suspend 3 mg of dried pellet in 300 μL of SDT buffer on ice for 2 h.
13. Centrifuge the samples at 13,000× *g* at 4 °C for 10 min.
14. Transfer the supernatant to new microcentrifuge tube.
15. Quantitate the protein concentration using a fluorescent assay of tryptophan content as modified from Wiśniewski et al [14] (*see* **Note 10**).
16. Immediately use protein samples for digestion or store at −80 °C.

3.2 Protein Digestion and Peptide Desalt

For protein digestion, the filter-aided sample preparation (FASP) method described by Wiśniewski et al. [14] is recommended.

1. Take 100 μg of protein from each sample and make up to a final volume of 300 μL with UA buffer (*see* **Note 3**).
2. Transfer protein sample to a 10 K MWCO, 0.5 mL Pierce™ Protein Concentrator (Thermo Scientific) and centrifuge at 10,000× *g* for 30 min.
3. Discard the flow-through and wash the Concentrator two times with UA buffer by centrifugation at 10,000× *g* for 30 min each time.
4. Add 100 μL of 50 mM iodoacetamide, vortex shortly, and incubate for 30 min at room temperature in the dark.
5. Centrifuge at 10,000× *g* for 15 min and wash samples with UA buffer three times by centrifuge at 10,000× *g* for 30 min each.
6. Add 300 μL of 50 mM NH_4HCO_3 solution to wash the concentrator two times.
7. Suspend the protein in the concentrator with 100 μL of NH_4HCO_3 solution by vortex.
8. Digest the protein with trypsin with the ratio of protein: trypsin at ~100:1 (w/w) for 16 h at 37 °C.
9. Centrifuge at 10,000× *g* for 15 min to collect the elutes with peptide mixtures.
10. Add 10% trifluoroacetic acid to a final concentration of 0.4% to stop the digestion and acidify the solution.
11. Desalt the peptide mixture on Pierce™ C-18 Spin columns (Thermo Scientific) (*see* **Note 11**).
12. Lyophilize the desalted peptide mixtures in a centrifugal speed vacuum concentrator (Eppendorf).
13. Dissolve peptide mixtures in 20 μL 0.1% formic acid for LC-MS/MS analysis or store them at −80 °C.

3.3 LC-MS/MS Analysis

Peptide mixtures are analyzed with an Orbitrap Fusion Lumos mass spectrometer (Thermo Scientific) coupled to a NanoLC-ESI system (*see* **Note 12**). Two runs as technical replicates are suggested for each sample.

1. Balance the C18 analysis column (Acclaim PepMap™ 100, 100 μm × 2 cm, Thermo Scientific) with solvent A (0.1% formic acid in water).
2. Load peptide mixtures onto a DIONEX UltiMate 3000 RSLCnano System (Thermo Scientific).
3. Separate the peptide mixtures in a linear gradient in 90 min from 0% to 35% solvent B (0.1% formic acid in acetonitrile) at a flow rate of 0.3 μL/min.

4. Operate the mass spectrometer in positive ionization mode.
5. Set full-scan mass spectra over a range of 350–1700 m/z at a resolution of 60,000 for MS and 15,000 for MS/MS.
6. Select the 20 most abundant precursor ions in each MS scan for higher-energy collisional dissociation fragmentation at HCD collision energy of 30%.

3.4 Protein Identification and Qualification

For Label-free quantitation proteomics, we suggest using the Proteome Discoverer software (version 2.3; Thermo Scientific) to identify proteins and quantitate proteins simultaneously (*see* **Note 13**).

1. Download plant species-specific protein database in fasta format and add to the software.
2. Search against both the related protein database and a decoy database containing the randomized sequences of the original database by SEQUEST algorithms.
3. Set the search parameters as below: trypsin is chosen as the with missed cleavages allowed. Carbamidomethylation modification at Cys is chosen as a fixed modification and oxidation at Met as a variable one. Precursor mass tolerance is set as 10 ppm, fragment mass tolerance is 0.02 Da. Workflow for both processing and post-processing steps is present in Fig. 2.

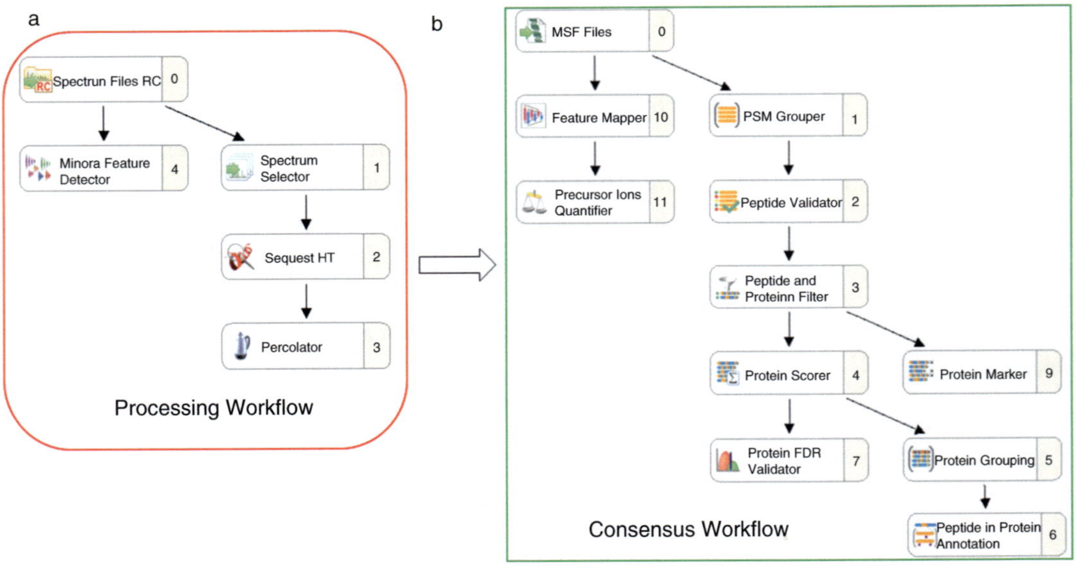

Fig. 2 The workflow of Label-free quantitation (LFQ) proteomics data processing. (**a**) Mass spectra raw data processing tree with the key nodes. Low quality mass spectra will be filtered out and the left ones will be searched against both the original protein database and the decoy database produced from the original one; the proteins with a false discovery rate less than 1% will be further analyzed. (**b**) Post-processing step (consensus step) for data analysis. Protein quantitation is completed by this step and peptide annotation as well as protein grouping are generated

4. For protein identification, the search results are passed through additional filters as follows: set peptide confidence greater than 95% and "Master protein" confidence greater than 99%, before exploring the data. A protein with FDR (false discovery rate) less than 1% is marked as "High" with experimental q-value no more than 0.01.
5. For protein quantitation, analyses are only focused on those confident proteins marked as "High" which must be found in at least one treatment.
6. Label-free quantitative methods integrated in Proteome Discoverer software is applied to quantitate proteins among different samples (*see* **Note 14**).

3.5 Bioinformatics

A protein with a fold change of more than two and the adjusted p-value less than 0.05 is selected as a differentially accumulated protein (DAP). For downstream DAP analyses normally include below items:

1. Venn Analysis and Heat Map (*see* **Note 15**).
2. Gene ontology (GO) annotation and enrichment analysis (*see* **Note 16**).
3. Kyoto Encyclopedia of Genes and Genomes (KEGG) pathways (*see* **Note 17**).
4. Protein and protein interaction network (*see* **Note 18**).

4 Notes

1. For β-mercaptoethanol containing solutions, β-mercaptoethanol should be added freshly before use, which is applied to PMSF containing solutions.
2. Freshly prepare the SDT buffer and add the protein inhibitor cocktail right before use.
3. For urea-containing buffer, both the handling and storing temperature should be below 30 °C.
4. Proteins dissolved in SDT buffer need not be reduced by DTT and can be directly alkylated by iodoacetamide.
5. Both solvents A and B should be sonicated at room temperature for 30 min to remove air bubbles.
6. Grinding the samples by the same person will increase the chances of reproducible protein yield.
7. For more than 1 g of fresh materials, using 20× volume of −20 °C stored TCA-Acetone solution will be suggested.
8. For large pellets, 20–30× volume of acetone solution is suggested to remove the salts.

9. Do not dry the pellets too much, otherwise it is much harder to dissolve the proteins.
10. To quantify protein samples, 0.5 μg/μL tryptophan solution stock solution will be first diluted into final serial concentrations of (ng/μL) 0, 25, 50, 100, 150, and 200 ng/μL in a 500 μL system. Next, add 3 μL of both the diluted tryptophan solutions or protein solutions to 1 ml dilution buffer and vortex to mix. Continue with 200 μL of the mixtures by transferring to a 96-well plate with three replicates each. The fluorescence signals are then measured at 295 nm for excitation and 350 nm for emission, with the slits set to 10 nm. A standard curve against tryptophan concentrations is drawn as the reference to calculate the sample protein concentration, which is the corresponding tryptophan concentration divided by 0.013.
11. Peptide desalting is carried out on Pierce™ C-18 Spin columns (Thermo Scientific). Use 200 μL of methanol to rinse the columns two times and 200 μL of equilibration buffer three times, followed by condition & elution buffer three times before the peptide mixtures are eluted by condition & elution buffer. Pool the desalted peptide mixtures and dry them in a centrifugal speed vacuum concentrator. Store dried peptide mixtures at −80 °C.
12. Dissolve the dried peptides mixture corresponding to 100 μg protein in 40 μL of 0.1% formic acid, resulting in 1~2 μg/μL peptide mixtures. We inject 1 μL for MS/MS analysis with 2-3 technique repeats each.
13. For each biological repeat, spectra from the technical repeats are combined into one file and search against the relevant plant species-specific protein database by SEQUEST algorithms. A decoy database created by randomizing the original protein sequences is also searched to calculate the false discovery rate (FDR). We set the FDR less than 1%.
14. Quantitate the protein-protein ratio (fold change) using a median of all possible pairwise ratios that are calculated between replicates of all peptide abundance pairs. We define a protein is a differentially accumulated protein if the fold change is more than 2 and the adjusted p-value is less than 0.05.
15. Use the R Packages of Venneuler and Pheatmap to generate the Venn diagram and heatmap, respectively.
16. Run the web tool of GENEONTOLOGY (http://geneontology.org/) with Fisher's test, and those of false discovery rate (FDR) corrected p-value or q-value less than 0.05 are selected as enriched GO terms. Visualization of the GO enrichment is completed by REVIGO (http://revigo.irb.hr/) with default parameters and ggplot2 package in R.

17. The online KOBAS is used to complete the Kyoto Encyclopedia of Genes and Genomes (KEGG) enrichment analysis of the differentially accumulated proteins.
18. Protein-protein interaction network is completed using the web tool of STRING (https://www. string-db.org).

Acknowledgments

This work was supported by National Nature Science Foundation of China (32070279,31370280), Natural Science Foundation of Jiangsu Province (BK20221560 and BK20141511), Project of Priority and Key Areas, Institute of Soil Science, Chinese Academy of Sciences (ISSASIP2206).

References

1. Kosova K, Vitamvas P, Prasil IT, Renaut J (2011) Plant proteome changes under abiotic stress – contribution of proteomics studies to understanding plant stress response. J Proteome 74:1301–1322
2. Ong SE, Blagoev B, Kratchmarova I, Kristensen DB, Steen H, Pandey A, Mann M (2002) Stable isotope labeling by amino acids in cell culture, SILAC, as a simple and accurate approach to expression proteomics. Mol Cell Proteomics 1:376–386
3. Gygi SP, Rist B, Gerber SA, Turecek F, Gelb MH, Aebersold R (1999) Quantitative analysis of complex protein mixtures using isotope-coded affinity tags. Nat Biotechnol 17:994–999
4. Ross PL, Huang YLN, Marchese JN, Williamson B, Parker K, Hattan S, Khainovski N, Pillai S, Dey S, Daniels S, Purkayastha S, Juhasz P, Martin S, Bartlet-Jones M, He F, Jacobson A, Pappin DJ (2004) Multiplexed protein quantitation in Saccharomyces cerevisiae using amine-reactive isobaric tagging reagents. Mol Cell Proteomics 3:1154–1169
5. Old WM, Meyer-Arendt K, Aveline-Wolf L, Pierce KG, Mendoza A, Sevinsky JR, Resing KA, Ahn NG (2005) Comparison of label-free methods for quantifying human proteins by shotgun proteomics. Mol Cell Proteomics 4: 1487–1502
6. Karim MR, Wang R, Zheng L, Dong X, Shen R, Lan P (2020) Physiological and proteomic dissection of the responses of two contrasting wheat genotypes to nitrogen deficiency. Int J Mol Sci 21
7. Yan M, Xue C, Xiong Y, Meng X, Li B, Shen R, Lan P (2020) Proteomic dissection of the similar and different responses of wheat to drought, salinity and submergence during seed germination. J Proteomics 220
8. Liu C, Niu G, Li X, Zhang H, Chen H, Hou D, Lan P, Hong Z (2021) Comparative label-free quantitative proteomics analysis reveals the essential roles of N-Glycans in salt tolerance by modulating protein abundance in Arabidopsis. Front Plant Sci 12
9. Yan M, Zheng L, Li B, Shen R, Lan P (2021) Comparative proteomics reveals new insights into the endosperm responses to drought, salinity and submergence in germinating wheat seeds. Plant Mol Biol 105:287–302
10. Li B, Zheng L, Wang R, Xue C, Shen R, Lan P (2022) A proteomic analysis of Arabidopsis ribosomal phosphoprotein P1A mutant. J Proteomics 262
11. Zhang X, Xue C, Wang R, Shen R, Lan P (2022) Physiological and proteomic dissection of the rice roots in response to iron deficiency and excess. J Proteomics 267
12. Lan P, Li W, Wen T-N, Shiau J-Y, Wu Y-C, Lin W, Schmidt W (2011) iTRAQ protein profile analysis of Arabidopsis roots reveals new aspects critical for iron homeostasis. Plant Physiol 155:821–834
13. Lan P, Li W, Schmidt W (2012) Complementary proteome and transcriptome profiling in phosphate-deficient Arabidopsis roots reveals multiple levels of gene regulation. Mol Cell Proteomics 11:1156–1166
14. Wisniewski JR, Zougman A, Nagaraj N, Mann M (2009) Universal sample preparation method for proteome analysis. Nat Methods 6:359–U360

Chapter 8

Chromatin Immunoprecipitation (ChIP) to Study the Transcriptional Regulatory Network that Controls Iron Homeostasis in *Arabidopsis thaliana*

Fei Gao and Christian Dubos

Abstract

In plants, gene expression is orchestrated by thousands of transcription factors (TFs). For instance, a large set of bHLH TFs are involved in the regulation of iron homeostasis in *Arabidopsis thaliana*. The identification of the direct target genes of TFs through uncovering the interaction between the TFs and *cis*-regulatory elements has become an essential step toward a comprehensive understanding of the iron homeostasis transcriptional regulatory network in Arabidopsis. Chromatin immunoprecipitation (ChIP) followed by qRT-PCR (ChIP-qPCR), sequencing (ChIP-seq), or chip hybridization (ChIP-chip) is a robust tool to investigate protein-DNA interactions in plants in a physiological context. The procedure generally includes six steps: DNA-protein crosslink, isolation of nuclei, shearing of chromatin, immunoprecipitation, DNA purification, and qRT-PCR analyses. In this protocol, we describe guidelines, experimental setup, and conditions for ChIP experiment in Arabidopsis. This protocol focuses on seedlings grown in control and iron deficiency conditions, but can readily be adapted for use with other Arabidopsis tissues or samples. In addition, the protocol could also be applied to perform ChIP-chip or ChIP-seq experiments.

Key words Arabidopsis, Transcription factor, *Cis*-element, Chip, Chromatin immunoprecipitation, Qrt-pcr

1 Introduction

Iron is one of the most important microelements for plant survival and proliferation [1]. On the other hand, excess iron is also toxic for plant growth due to the generation of reactive oxygen species through the Fenton reaction. Thus, the iron levels in plant cells must be fine-tuned to maintain iron homeostasis. At the transcriptional level, genes involved in iron uptake, translocation, assimilation, and storage are tightly regulated by a set of transcription factors (TFs) among which bHLH play a central role [2–7]. Such regulations rely on the direct interaction between TFs and specific DNA sequences known as *cis*-regulatory elements usually localized

Jeeyon Jeong (ed.), *Plant Iron Homeostasis: Methods and Protocols*, Methods in Molecular Biology, vol. 2665,
https://doi.org/10.1007/978-1-0716-3183-6_8,

at the promoter region of the target genes [8–10]. Therefore, the determination of interactions between TFs and *cis*-regulatory sequences is an essential step toward a comprehensive understanding of the transcriptional regulatory code that controls several facets of plant growth and development, including the regulation of iron homeostasis.

The interactions between TFs and *cis*-regulatory elements have been widely studied through various approaches. Several *in vitro* techniques, such as yeast one-hybrid (Y1H) assays, electrophoretic mobility shift assays (EMSA), and DNA pull-down assays are commonly used to determine protein-DNA interactions [10]. However, in vitro interactions could be forced and might fail to replicate the in vivo situation occurring in plant cells [10]. Chromatin immunoprecipitation (ChIP) is an antibody-based technology used to selectively enrich specific DNA-binding proteins along with their DNA targets, which become a powerful tool to detect in vivo (i.e., cells, tissues or seedlings) protein-DNA interactions [11]. ChIP coupled with quantitative PCR (ChIP-qPCR) is employed to determine the direct association between TFs and *cis*-regulatory elements. Furthermore, ChIP coupled with quantitative sequencing (ChIP-seq) or DNA-chip was used to identify the direct target genes of several TFs. In addition, this approach allows investigating changes in protein association intensity with target DNA in response to different environmental conditions. For example, two studies demonstrated in Arabidopsis that bHLH121 could directly associate with the promoter region of Ib bHLH TFs to activate their expression in response to iron deficiency using ChIP-qPCR [5, 12]. In addition, it was also shown using ChIP-seq that the binding of bHLH121 to the promoter of subgroup Ib genes is increased under iron deficiency when compared to the control condition [13].

In this chapter, we describe the guidelines, experimental setup, and conditions for ChIP-qPCR experiment in Arabidopsis. This protocol was designed to compare seedlings grown under control and iron deficiency conditions but can readily be adapted for use with other Arabidopsis tissues or samples of plants grown under different conditions. In addition, the protocol described herein could also be applied to perform ChIP-chip or ChIP-seq experiments.

2 Materials

2.1 Plant Materials and Media

Plant materials:

(a) A transgenic *Arabidopsis thaliana* line expressing *probHLH121:gbHLH121-GFP* in *bhlh121* mutant background generated by Gao et al. [5].

(b) An Arabidopsis transgenic line expressing free GFP.

Plant media:

Half-strength Murashige and Skoog (MS) medium: half-strength MS salts, 0.05% (w/v) MES, 1% (w/v) sucrose, and 0.7% (w/v) agar. Add 50 μM Fe(III)-EDTA for the control medium and 0 μM Fe for the Fe deficiency medium.

2.2 Other Reagents

1. ChIP-grade anti-GFP antibody (Abcam).
2. Dynabeads® protein G (Thermo Fisher Scientific).
3. TB Green Premix Ex Taq II (Takara).
4. 10 mg/ml Proteinase K
5. Phenol:chloroform:isoamyl alcohol (25:24:1, pH 8.0).

2.3 Solutions

All solutions are prepared with DNase-free water.

1. Fixation buffer: 1% formaldehyde. Dilute from 37% formaldehyde stock. Must be prepared freshly (**Note 1**).
2. 2 M glycine
3. Nuclei isolation buffer (NIB): 20 mM PIPES-KOH (pH 7.6), 1 M Hexylene glycol, 10 mM $MgCl_2$ 500 μl, 0.1 mM EGTA, 15 mM NaCl, 60 mM KCl, 0.5% Triton X100, 5 mM β-mercaptoethanol, 1x Roche cOmplete protease inhibitor (EDTA-free), pre-chill before use.
4. Nuclei lysis buffer (NLB): 50 mM Tris-HCl (pH 8), 10 mM EDTA, 1% SDS, 1x Roche cOmplete protease inhibitor (EDTA-free), pre-chill before use.
5. ChIP dilution buffer (CDB): 1.1% Triton X100, 1.2 mM EDTA, 16.7 mM Tris-HCl (pH 8), 167 mM NaCl, pre-chill before use.
6. Low salt wash buffer (LSWB): 150 mM NaCl, 0.1% SDS, 1% Triton X10, 20 mM Tris-HCl (pH 8), 2 mM EDTA, pre-chill before use.
7. High salt wash buffer (HSWB): 150 mM NaCl, 0.1% SDS, 1% Triton X10, 20 mM Tris-HCl (pH 8), 2 mM EDTA, pre-chill before use.
8. LiCl wash buffer (LiWB): 0.25 M LiCl, 1% NP40 (IGEPA CA-630), 1% Sodium deoxycholate, 1 mM EDTA, 10 mM Tris-HCl (pH 8), pre-chill before use.8. TE wash buffer (TE): 10 mM Tris-HCl (pH 8), 1 mM EDTA, pre-chill before use.
9. Elution buffer (EB): 1% SDS, 0.1 M $NaHCO_3$.

2.4 Labware and Equipment

1. 1.5 ml Protein LoBind Tubes (Eppendorf)
2. 50 ml falcon tubes
3. DNase-free pipette tips.
4. 40 μm/70 μm Mesh Filter Cell Strainers (Biologix)
5. Vacuum pump.
6. Bioruptor.
7. Rotating mixer.
8. 65 °C incubator
9. 4 °C cold room
10. PCR machine and Real-time PCR machine.
11. Magnetic rack.
12. Refrigerated centrifuge.
13. Mortar and pestle.

3 Methods

3.1 Plant Growth and Treatment

1. Surface-sterilize and sow Arabidopsis seeds on 1/2 MS agar medium containing 50 μM Fe in 12 × 12 cm square Petri dishes (Fig. 1; **Note 2**). Put the dishes in the growth chamber at 22 ° C under long-day conditions for 7 days (16 h light / 8 h dark cycle; light intensity: 120 μmol/cm^2/s).
2. Transfer the 7-day-old seedlings to 1/2 MS agar medium containing 0 μM Fe (iron deficiency) or 50 μM Fe (control), and put them under long-day conditions for another 3 days (Fig. 1; **Note 2**).

3.2 Seedlings Collection and Fixation

1. Harvest 2 *g* of seedlings into a 50 ml falcon tube containing 30 ml of fixation buffer (Fig. 1; **Note 1**).
2. Cross-link seedlings with fixation buffer under vacuum infiltration for 15 min.

   Break and re-apply vacuum after 5 and 10 min to facilitate penetration of formaldehyde into the seedlings.
3. Stop the cross-link by adding fresh 2 M glycine to every 10 ml of fixation buffer to a final concentration of 0.125 M. Apply vacuum for another 5 min.
4. Remove the glycine and rinse the samples three times with Milli-Q water. Remove excess liquid from samples, and dry samples using KIMTECH paper.
5. Snap freeze sample in liquid nitrogen. Store at −80 °C (**Note 3**).

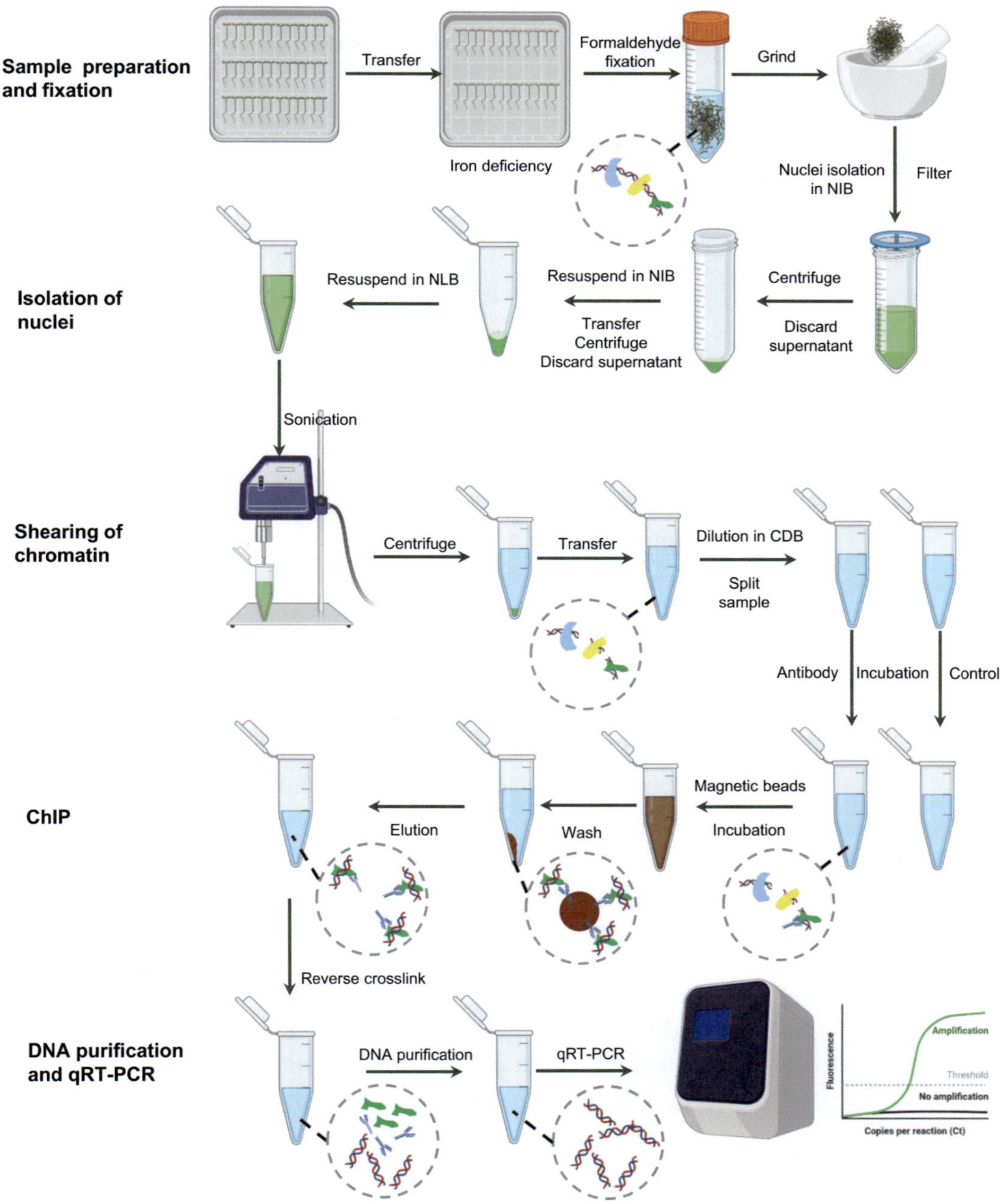

Fig. 1 Outline of the ChIP-qPCR procedure. After cross-linking with formaldehyde, the cells are lysed, and the nuclei-containing fraction is isolated. After chromatin shearing, the samples are incubated with antibodies for immunoprecipitation. After crosslink reversal, PCR-ready DNA is collected. Binding of the protein of interest to genomic DNA is examined by qRT-PCR

3.3 Isolation of Nuclei and Shearing of Chromatin

1. Grind the cross-linked samples to a fine powder using a mortar and a pestle in liquid nitrogen (Fig. 1; **Note 4**).

2. Add 7 ml of NIB to the powder and let it on ice for 15 min until the sample is thawed. Transfer the sample to a 50 mL conical tube (**Note 5**).
3. Filter the extract through pre-wet 70 μm mesh filter cell strainers, and then through pre-wet 40 μm mesh filter cell strainers into a fresh 50 ml falcon tube. Wash the strainers and tubes using NIB, and adjust the volume of extract to 8 ml (Fig. 1).
4. Centrifuge filtrate at 1500 *g* for 15 min. Discard supernatant.
5. Resuspend the pellet gently and thoroughly with 1 ml of NIB. Transfer the sample to a 1.5 ml microcentrifuge tube (Fig. 1).
6. Centrifuge filtrate at 1500 *g* for 15 min. Discard supernatant.
7. Resuspend the pellet gently and thoroughly with 220 μl of NLB and let it on ice for 5 min.
8. Sonicate the chromatin solution using Bioruptor on the ice for 20 cycles of 30 s ON and 30 s OFF at the HIGH setting. Mix samples between each 5 min period (Fig. 1; **Note 6**).
9. Centrifuge sonicated chromatin at maximum speed for 10 min at 4 °C. Transfer the clean supernatant to a new 1.5 ml Eppendorf tube.
10. Repeat **step 16** one more time to remove as much of the pellet as possible. The typical yield after sonication is 220 μl.
11. Take 10 μl of nuclear lysate to check sonication efficiency. And freeze 10 μl nuclear lysate as an input sample.

3.4 Pre-clearing

1. Dilute the sonicated nuclear lysate ten-fold with CDB.
2. For pre-clearing, prepare a batch of magnetic protein G beads (5 μL per tube). Wash the beads with 1 ml of CDB three times. Discard the solution using a magnetic rack. Resuspend the beads in their initial volume using CDB.
3. Add 5 μl of magnetic protein G beads per tube. Incubate for 2 h at 4 °C with gentle agitation on a Rotating mixer.
4. Transfer the supernatant to a new 1.5 mL Eppendorf tube using a magnetic rack.
5. Split the chromatin solution into 2 Eppendorf tubes (one for no antibody control and one for immunoprecipitation).

3.5 ChIP and Reverse Crosslink

1. Add 5 μg anti-GFP antibody per IP tube. Incubate for 4 h to 8 h at 4 °C with gentle agitation on a rotating mixer (Fig. 1; **Note 7**).
2. Prepare a batch of magnetic protein G beads (35 μl per tube) as in **step 20**.

3. Add 35 μl of beads into each chromatin tube (control and IP). Incubate at 4 °C with gentle agitation on a rotating mixer for at least 2 h.
4. Collect the immune complexes using a magnetic rack. Discard supernatant.
5. Wash beads for a total of eight rounds with 1 ml of the following buffers. Collect beads on a magnetic rack and remove supernatant after each round of wash (Fig. 1).
   (a) One round of 5-min wash on the rotating mixer and one round of quick wash with Low Salt Wash Buffer (LSWB).
   (b) One round of 5-min wash on the rotating mixer and one round of quick wash with High Salt Wash Buffer (HSWB).
   (c) One round of 5-min wash on the rotating mixer and one round of quick wash with LiCl Wash Buffer (LiWB).
   (d) One round of 5-min wash on the rotating mixer and one round of quick wash with TE buffer.
6. Resuspend immune complexes with 260 μl of Elution Buffer (EB). Incubate tubes at 65 °C for 15 min to the elute protein-DNA complex. Collect beads using magnetic rack. Transfer the supernatant to a fresh 1.5 ml Eppendorf tube.
7. Take the input sample from **step 18** out of the freezer. Add 250 μl of Elution Buffer (EB).
8. Add 10 μl of 5 M NaCl to both Input and ChIP samples. Reverse crosslinks overnight at 65 °C (Fig. 1).

3.6 DNA Purification

1. Add 6.5 μl of 0.5 M EDTA, 13 μl of 1 M Tris-HCl (pH 6.5), and 1.5 μl proteinase K. Incubate tubes at 42 °C for 2 h.
2. Add an equal volume (281 μl) of phenol/chloroform/isoamyl alcohol (25:24:1, pH 8.0). Vortex for 30 s and centrifuge at full speed for 10 min at room temperature.
3. Transfer 240 μl of supernatant into a new 1.5 ml Eppendorf tube.
4. To precipitate DNA, add 0.1 volume (24 μl) of 3 M NaAc and 2.5 volumes of ethanol (600 μl). Incubate at −80 °C for 20 min. Centrifuge at full speed at 4 °C for 20 min.
5. Remove the supernatant and wash the pellet with 500 μl ice-cold 70% ethanol. Centrifuge at full speed at 4 °C for 10 min.
6. Remove the supernatant and dry the pellet. Resuspend the pellet in 50 μl of nuclease-free water (**Note 8**).

3.7 Quantitative PCR and Data nalysis

1. Proceed to quantitative PCR using gene-specific primers (**Note 9**).
   - qPCR is carried out in a final volume of 10 μl containing: 5 μl of TB Green Premix Ex Taq II (2×), 1 μl of ChIP DNA template from **step 37,** 1 μl of mixed primers (forward and reverse, 5 μM each) and 3 μl of sterile Milli-Q water.
   - The amplification program used is the following: denaturation (95 °C, 5 min), then 45 cycles of denaturation (95 °C, 15 s), primer annealing (55 to 60 °C, 10 s), and extension (72 °C, 10 s).
   - At least three technical replicates are needed for each sample. Three independent biological replicate ChIP experiments should be performed.
2. Record the threshold cycles (Ct values) from the exponential phase of the qPCR for the IP and input DNA sample for each primer pair.
3. Calculate the DNA binding ratio (the relative amount of immunoprecipitated DNA compared to INPUT DNA for the control regions) using the following formula: DNA binding ratio $= 2^{-[Ct(IP)-\ 3.32-\ Ct(Input)]} * 100$ (3.32 is a compensatory factor to correct the input dilution).
4. Analyze the statistical differences between the experiment (bHLH121-GFP) and control samples (free GFP) at each target region using one-way ANOVA with the post hoc Tukey test. $P < 0.05$ is considered statistically significant (Fig. 2).

4 Notes

1. The formaldehyde solution needs to be fresh. Formaldehyde is highly toxic, work in a fume hood. The seedlings need to be submerged in the fixation solution before starting the vacuum infiltration and throughout the infiltration.
2. Put the dishes vertically to keep the whole seedling on the agar surface, which will help with seedling collection.
3. Sample can be stored at −80 °C for up to a few weeks.
4. Keep the samples frozen throughout the process.
5. Add the β-mercaptoethanol into the Nuclei Isolation Buffer before use. The β-mercaptoethanol is toxic, work in a fume hood.
6. Try to avoid foaming, the formation of bubbles could reduce the sonication efficiency. If samples get too hot, the proteins will be denatured. Keep the samples on the ice.
7. This step can also be done overnight.

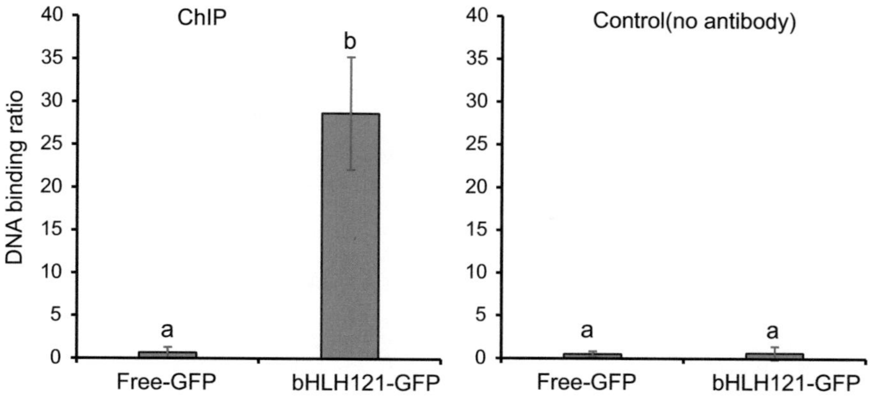

Fig. 2 Example of a ChIP-qPCR assay. Seven-day-old *Arabidopsis thaliana* seedlings carrying bHLH121-GFP or free GFP construct were exposed to iron deficiency treatment for 3 days under long day conditions. Anti-GFP antibody was used to perform the immunoprecipitation. The ChIP DNA was used as template for qPCR analysis, and the no antibody treatment DNA was used as negative control. The figure shows the DNA binding ratio (the relative amount of immunoprecipitated DNA compared to input DNA after qPCR analysis). Means within each condition with the same letter are not significantly different according to one-way ANOVA followed by post-hoc Tukey test, $p < 0.05$ ($n = 4$ to 8 technical repeats). Error bars show ±SD

8. The purified ChIP DNA can be stored at −20 °C for at least one month. Do not dilute before storage.
9. For ChIP-qPCR primer design, locate the potential gene regions your protein is bound to. PCR product size should be 70 to 150 bp long. Test the primers on genomic DNA for efficient amplification before using them on ChIP reactions.

Acknowledgments

FG was supported by a fellowship from the China Scholarship Council (CSC).

References

1. Briat J-F, Dubos C, Gaymard F (2015) Iron nutrition, biomass production, and plant product quality. Trends Plant Sci 20:33–40. https://doi.org/10.1016/j.tplants.2014.07.005
2. Gao F, Robe K, Gaymard F et al (2019) The Transcriptional Control of Iron Homeostasis in Plants: A Tale of bHLH Transcription Factors? Front Plant Sci 10:6. https://doi.org/10.3389/fpls.2019.00006
3. Tissot N, Robe K, Gao F et al (2019) Transcriptional integration of the responses to iron availability in Arabidopsis by the bHLH factor ILR3. New Phytol 223:1433–1446. https://doi.org/10.1111/nph.15753
4. Gao F, Dubos C (2021) Transcriptional integration of plant responses to iron availability. J Exp Bot 72:2056–2070. https://doi.org/10.1093/jxb/eraa556
5. Gao F, Robe K, Bettembourg M et al (2020) The Transcription Factor bHLH121 Interacts with bHLH105 (ILR3) and Its Closest Homologs to Regulate Iron Homeostasis in Arabidopsis. Plant Cell 32:508–524. https://doi.org/10.1105/tpc.19.00541

6. Gao F, Robe K, Dubos C (2020) Further insights into the role of bHLH121 in the regulation of iron homeostasis in Arabidopsis thaliana. Plant Signal Behav 1795582. https://doi.org/10.1080/15592324.2020.1795582
7. Grant-Grant S, Schaffhauser M, Baeza-Gonzalez P et al (2022) B3 transcription factors determine iron distribution and FERRITIN gene expression in embryo but do not control total seed iron content. Front Plant Sci 13:870078. https://doi.org/10.3389/fpls.2022.870078
8. Thévenin J, Xu W, Vaisman L et al (2016) The Physcomitrella patens system for transient gene expression assays. Methods Mol Biol 1482:151–161. https://doi.org/10.1007/978-1-4939-6396-6_10
9. Schmitz RJ, Grotewold E, Stam M (2022) Cis-regulatory sequences in plants: their importance, discovery, and future challenges. Plant Cell 34:718–741. https://doi.org/10.1093/plcell/koab281
10. Yamaguchi N, Winter CM, Wu M-F et al (2014) PROTOCOLS: Chromatin Immunoprecipitation from Arabidopsis Tissues. Arabidopsis Book 12:e0170. https://doi.org/10.1199/tab.0170
11. Lee JH, Jin S, Kim SY et al (2017) A fast, efficient chromatin immunoprecipitation method for studying protein-DNA binding in Arabidopsis mesophyll protoplasts. Plant Methods 13:42. https://doi.org/10.1186/s13007-017-0192-4
12. Lei R, Li Y, Cai Y et al (2020) bHLH121 Functions as a Direct Link that Facilitates the Activation of FIT by bHLH IVc Transcription Factors for Maintaining Fe Homeostasis in Arabidopsis. Mol Plant 13:634–649. https://doi.org/10.1016/j.molp.2020.01.006
13. Kim SA, LaCroix IS, Gerber SA, Guerinot ML (2019) The iron deficiency response in Arabidopsis thaliana requires the phosphorylated transcription factor URI. Proc Natl Acad Sci U S A 116:24933–24942. https://doi.org/10.1073/pnas.1916892116

Chapter 9

Comprehensive Survey of ChIP-Seq Datasets to Identify Candidate Iron Homeostasis Genes Regulated by Chromatin Modifications

Yang Yu, Yuxin Wang, Zhujun Yao, Ziqin Wang, Zijun Xia, and Joohyun Lee

Abstract

Vital biochemical reactions including photosynthesis to respiration require iron, which should be tightly regulated. Although increasing evidence reveals the importance of epigenetic regulation in gene expression and signaling, the role of histone modifications and chromatin remodeling in plant iron homeostasis is not well understood. In this study, we surveyed publicly available ChIP-seq datasets of Arabidopsis wild-type and mutants defective in key enzymes of histone modification and chromatin remodeling and compared the deposition of epigenetic marks on loci of genes involved in iron regulation. Based on the analysis, we compiled a comprehensive list of iron homeostasis genes with differential enrichment of various histone modifications. This report will provide a resource for future studies to investigate epigenetic regulatory mechanisms of iron homeostasis in plants.

Key words *Arabidopsis*, Iron homeostasis, Histone modification, Chromatin remodeling, ChIP-Seq

1 Introduction

Chromatin remodeling and histone modifications are associated with the condensation of chromatins that influences the accessibility of RNA polymerases and transcription factors [1]. Furthermore, transcriptional activity can be maintained mitotically and/or meiotically without changes in the nucleotide sequence via histone modification and chromatin remodeling [1]. An increasing number of epigenetic studies have revealed that chromatin status is controlled via environmental cues [2–9]. In *Arabidopsis thaliana*, epigenetic modifications and consequent changes in gene expression

Authors Yang Yu, Yuxin Wang, Zhujun Yao and Ziqin Wang have contributed equally to this chapter.

Jeeyon Jeong (ed.), *Plant Iron Homeostasis: Methods and Protocols*, Methods in Molecular Biology, vol. 2665,
https://doi.org/10.1007/978-1-0716-3183-6_9,

occur in response to changes in environmental conditions, including the availability of essential nutrients [10, 11].

While various aspects of iron homeostasis have been extensively studied, how iron homeostasis is affected by histone modifications or chromatin remodeling in plants has still not been well understood. Under iron-sufficient conditions, the symmetric demethylation of histone H4R3 (H4R3sme2) promotes transcriptional suppression of *basic helix-loop-helix (bHLH) 38*, *bHLH39*, *bHLH100*, and *bHLH101*, which encode interacting partners of FIT1 (FER-LIKE IRON DEFICIENCY INDUCED TRANSCRIPTION FACTOR 1) in Arabidopsis [12]. PRC2 (polycomb repressive complex 2)-mediated H3K27me3 was found to be deposited on *FIT*, *FRO2 (FERRIC REDUCTION OXIDASE 2)*, and *IRT1 (IRON-REGULATED TRANSPORTER 1)* loci and attenuate the expression of these FIT-dependent genes when they are induced under iron-deficient conditions [13]. A few additional studies have shown that chromatin remodeling contributes to regulation of iron in plants [14], but more efforts are required to uncover how chromatin modifications and epigenetic regulation affect iron homeostasis.

In this report, we surveyed histone modifications and histone variant depositions on iron homeostasis genes by comparing ChIP-Seq datasets that are publicly available from The Gene Expression Omnibus (GEO) [15] or Sequence Read Archive (SRA) [16]. To reduce the chances of including false-positive targets from ChIP-Seq datasets, which is a common issue in ChIP-seq analysis [17, 18], we compared ChIP-seq enrichment profiles from wild type and loss of function mutants of direct histone modifiers, such as histone methyltransferases, histone acetyltransferases, and chromatin remodelers. Using these comparative approaches, we compiled a comprehensive list of histone modifications on various iron homeostasis loci that can serve as a resource between iron homeostasis and epigenetics research.

2 Comparative Survey of ChIP-Seq Datasets between Wild Type and Epigenetic Mutants

Positively charged amino acids at the histone N-terminal tail are subjected to methylation or acetylation [19–21]. The acetyl group neutralizes the positively charged amino acid and loosens the interaction between histone and the negatively charged DNA, subsequently allowing more space for the transcriptional machinery [22]. Meanwhile, histone methylations can have either inductive or repressive effects on target gene expression by encompassing or detaching histone tails from the DNA [23]. In general, methylation on H3K4, K36, and K79 is associated with active transcription, and methylation on H3K9, K27, and H4K20 correlates with heterochromatin formation and gene repression [24–28].

To survey publicly available ChIP-seq datasets of selected epigenetic mutants and compare the epigenetic mark depositions in the mutants and wild type, we obtained the raw sequencing reads from GEO and SRA and first conducted a quality check FastQC [29]. We then processed the reads using Trimmomatic v0.39 to trim the adapter sequence and remove low-quality reads [30], and aligned the reads to the *Arabidopsis thaliana* genome (assembly TAIR10) using Bowtie v 2.3.5.1 [31]. The overall mapping rates were above 70% for all datasets. Samtools v1.6 was then used to convert, sort, and filter the mapped reads [32], and only the reads that mapped uniquely to the genome were kept. Biological replications were merged into one BAM file. Peak calling was performed with MACS v3.0.0 [33] using the parameters that m-fold = 5 to 50, broad-cutoff = 0.1, and *P* value cutoff = 1e-05. Differential peak analysis was performed using DiffBind v.3.16 [34]. Peaks were annotated to genes when located around 1000 bp from the transcription start site (TSS) using ChIPSeeker [35]. Only the genes that show significant enrichment were regarded as differentially modified by each histone modification or remodeling factor (Fig. 1).

3 H3K4me2 and H3K4me3 Depositions via ATX3, ATX4, and ATX5

Histone 3 lysine 4 dimethylation (H3K4me2) and histone 3 lysine 4 trimethylation (H3K4me3) are considered to be involved in activating gene expression [36, 37]. ARABIDOPSIS TRITHORAX (ATX) proteins are histone methyltransferases identified as homologs of the TRITHORAX GROUP (trxG) in *Drosophila*, and regulate multiple biological processes, including floral organ development, fiber secondary wall synthesis, and drought stress in Arabidopsis [38–41]. Among 12 ATX family members in Arabidopsis, five are categorized as the ATX group, and the other seven are classified as ATX-related (ATXR) group [42]. Within the ATX subgroup, ATX3, ATX4, and ATX5 are the closest homologs, and ChIP-Seq analysis revealed that the *atx3 atx4 atx5* has decreased H3K4me2 and H3K4me3 depositions genome-wide [42, 43]. These data suggest that ATX3, ATX4, and ATX5 play redundant roles in regulating H3K4me2 and H3K4me3 [42].

To compare H3K4me2 and H3K4me3 depositions on iron homeostasis gene loci, we conducted the analysis described in the previous section using a ChIP-seq dataset generated from wild type and the *atx3 atx4 atx5* triple mutants, GSE83802 from GEO (Fig. 1; Table 1). In the H3K4me2 ChIP-Seq data, *MEMBRANE PROTEIN OF ER BODY 1* (*MEB1*) was the only gene with significantly less H3K4me2 depositions in the mutant among iron

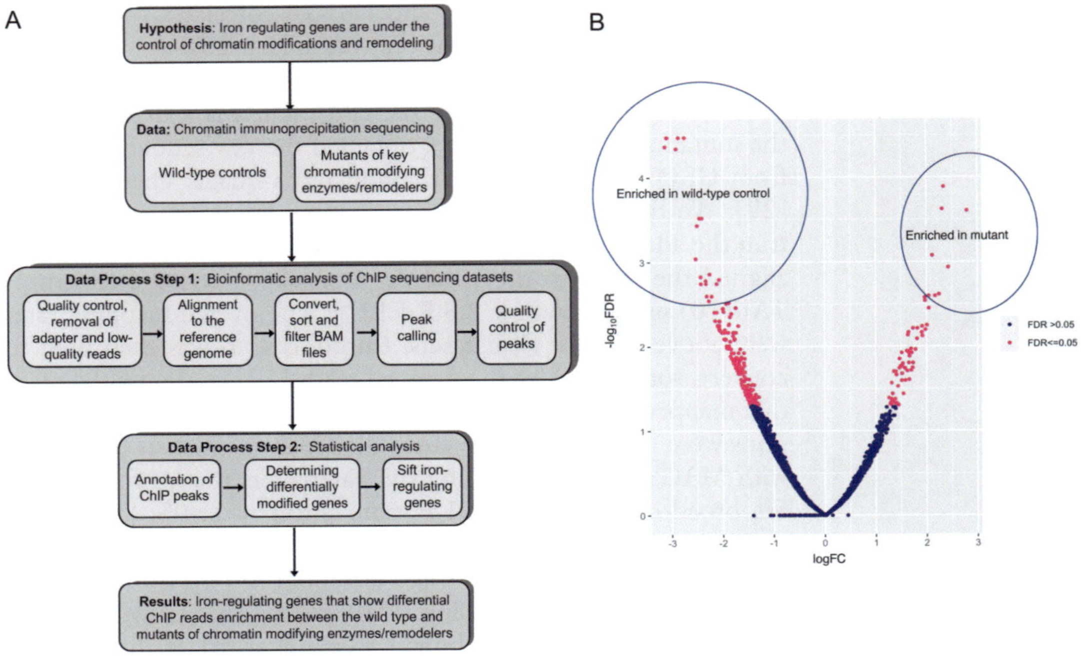

Fig. 1 Summary of the research strategy in this study. (**a**) Flowchart showing the research strategy and methods. (**b**) Volcano plot demonstrating the criteria to identify the differentially modified genes. Log fold change (logFC) is measured by the logarithm of the fold change of differential peaks in the mutant versus the control. Peaks with false discovery rate (FDR) smaller or equal to 0.05 and logFC greater than 2 are clustered as "enriched in mutant." Peaks with false discovery rate (FDR) smaller or equal to 0.05 and logFC smaller than −2 are clustered as "enriched in wild-type control"

homeostasis genes (Fig. 3). MEB1 is a vacuolar iron transporter (VIT) family member localized to the ER membrane and heterologous expression in yeast showed that it transports iron, suggesting that MEB1 may sequester iron from cytosol to the ER for iron detoxification [44]. Significant changes in H3K4me3 deposition were detected on the locus of *IRON-REGULATED TRANSPORTER-1/FERROPORTIN1* (*IREG1/FPN1*), which encodes an iron effluxer on the plasma membrane [45], and in *YELLOW STRIPE-LIKE 3* (*YSL3*), which encodes an iron-nicotianamine transporter required for long-distance signaling with *YELLOW STRIPE-LIKE1* (*YSL1*) [46]. Although H3K4me3 deposition on *IREG1/FPN1* was decreased in *atx3 atx4 atx5* compared to the deposition in the wild type, H3K4me3 deposition on *YSL3* was increased in the mutant, suggesting that the *YSL3* locus is not a direct target of ATX3, ATX4, or ATX5 (Fig. 3).

Table 1
Summary of iron homeostasis-related genes that were found to exhibit differential deposition of histone modifications between wild type and the mutant

Chromatin modification	Mutated gene(s) (ID)	Source of dataset	Accession number	Differentially enriched iron homeostasis-related genes
H3K4me2	*ATX3 ATX4 ATX5* (AT3G61740, AT4G27910, AT5G53430)	[42]	GSE83802	*MEB1*
H3K4me3	*ATX3 ATX4 ATX5* (AT3G61740, AT4G27910, AT5G53430)			*YSL3, IREG1/FPN1*
H3K27me3	*CLF* (AT2G23380)	[58]	GSE108960	*FRO2, VTL1, VTL2, VIT1, ISU2, FRD3, S8H, CYP82C4, ELS1, IREG3/FPN3, FEP3/IMA1, VTL5, MTP8, ISU3, FRO5, FRO8*
	SWN (AT4G02020)			*FRO1, FRO2, VTL1, FEP3/IMA1, FRD3, MTP8, ISU3, IRT1, FRO8*
H3K36me3	*SDG8* (AT1G77300)	[72]	SRP149810	*AHA2, FRO7, IRT3, NRAMP1, OPT3, YSL1, YSL3, VIT, FPN3, FER1, FER3, bHLH34, bHLH115, PYE*
H3ac	*HAC1 HAC5* (AT1G79000, AT3G12980)	[75]	GSE162459	*BCD1, BTS, PYE*
H3Th3ph	*AEL3 AEL4* (AT3G03940, AT5G18190)	[83]	GSE68370	*IRT2, IRT3, FEP3/IMA1*
H2A.Z	*PIE1* (AT3G12810)	[91]	GSE139459	*FRO2, MIT1, MIT2, BTSL1, BTSL2, SCD1, IRT3, VTL2, VTL5, FIT, ZIP3, FEP1/IMA3, FER2, FER3, FER4, FRD3, S8H, F6H1, BTS, PDR9, OPT3, IRT1, IRT2, ISU1, MEB1, MEB2, AHA2, FPN2, FPN3, HSCB, AT12CYS-2, ELS1, YSL2*

The list includes genes that encode: VIT1, VTL1, VTL2, VTL: vacuolar iron transporter family members [95, 107]; ISU1, ISU3: iron-sulfur cluster assembly enzymes [108]; S8H: a coumarin biosynthesis enzyme [109]; CYP82C4: a cytochrome P450 enzyme [110]; ELS1: a MATE transporter for iron remobilization [111]; IREG1/FPN1: an iron effluxer on the plasma membrane [45]; FRO1, FRO2, FRO5, FRO7, FRO8: ferric chelate reductases [112, 113]; FEP3/IMA1: an iron uptake inducing peptide [114]; FRD3: a xylem citrate effluxer for efficient iron translocation [115, 116]; MTP8: a vacuolar iron/manganese transporter [117]; NRAMP1: a high affinity manganese transporter [118, 119]; OPT3: a transporter that loads iron into the phloem and plays a role in shoot-to-root signaling [120]; YSL1, YSL2, YSL3: transporters of iron-nicotianamine complexes [121]; FER3: an iron sequestering protein [122]; IRT3: a plasma membrane localized zinc/iron transporter [123];bHLH115: a basic helix-loop-helix transcription factors involved in iron homeostasis [124]; BCD1: a Golgi-localized transporter involved in iron homeostasis under osmotic stress [100, 125]; BTS: a E3 ligase and negative regulator of iron deficiency response [126, 127]; PYE: a regulator of iron deficiency response and BTS binding partner [128]; ISU1: a regulator of mitochondrial iron homeostasis [102, 129]; MEB1, MEB2: putative iron/metal transporters [44]; IMA3/FEP1: small peptide that activates iron deficiency response genes [93, 130, 131]; FIT: a positive regulator of iron uptake genes under iron deficiency [104]; BTSL1, BTSL2: BTS-like E3 ligases and negative regulators of iron deficiency response [106, 132]

4 H3K27me3 Deposition via CLF or SWN

In higher eukaryotes, H3K27me3 is a representative silencing mark established by Polycomb repressive complex 2 (PRC2) [47, 48]. Among PCR2 components in Arabidopsis, CURLY LEAF (CLF), SWINGER (SWN), and MEDEA (MEA) are paralogs of Enhancer of zeste (E(z)) in *Drosophila*, which trimethylates H3K27 [49]. MEA is shown to specifically regulate seed development and reproduction, while CLF and SWN participate in general plant development regulation such as embryo, root, leaf, and flower development [50–53]. CLF has been reported to affect iron deficiency response in Arabidopsis [13, 54]. In the *clf* mutant, the enrichment of H3K27me3 was decreased in iron acquisition genes, *FIT*, *FRO2*, and *IRT1*, compared to that of wild-type in Arabidopsis roots under both iron-sufficient and -deficient conditions and attenuated induction of FIT-dependent iron deficiency response genes. Although CLF and SWN share a redundant role, they also have distinguishable targets depending on tissues and developmental stages [53, 55–59]. Therefore, we separately analyzed the datasets obtained with studies on CLF and SWN using the ChIP-Seq data from GSE108960 [58]. As shown in Table 1, the comparative analysis between wild type and *clf* mutant, differential H3K27me3 deposition was identified on 16 loci of iron homeostasis-related genes, including *FRO2*, *FRO5*, *FRO8*, *VIT1*, *VTL1*, *VTL2*, *VTL5*, *ISU2*, *ISU3*, *FERRIC REDUCTASE DEFECTIVE3* (*FRD3*), *SCOPOLETIN 8- HYDROXYLASE* (*S8H*), *CYP82C4*, *ESD4-LIKE SUMO PROTEASE 1* (*ELS1*), *IRON-REGULATED PROTEIN 1* (*IREG1/FPN1*), *FE-UPTAKE-INDUCING PEPTIDE3/IRONMAN 1* (*FEP3/IMA1*), and *METAL TOLERANCE PROTEIN8* (*MTP8*) (Fig. 2; Table 1). From the comparative analysis of wild type and *swn* mutant ChIP-seq datasets, loci of iron homeostasis genes such as *FRO1*, *FRO2*, *VTL1*, *FEP3/IMA1*, *FRD3*, *MTP8*, *ISU3*, *IRT1*, and *FRO8* were found to be associated with differential H3K27me3 deposition (Table 1). The genes detected from the *clf* and *swn* datasets largely overlap, suggesting that CLF and SWN are likely to play redundant role(s) in regulating iron homeostasis genes. Among the 18 E(z)-mediated iron regulatory genes identified from our analysis (Table 1), *FRO2, IRT1,* and *FRD3* have also been identified as H3K27me3 targets [54].

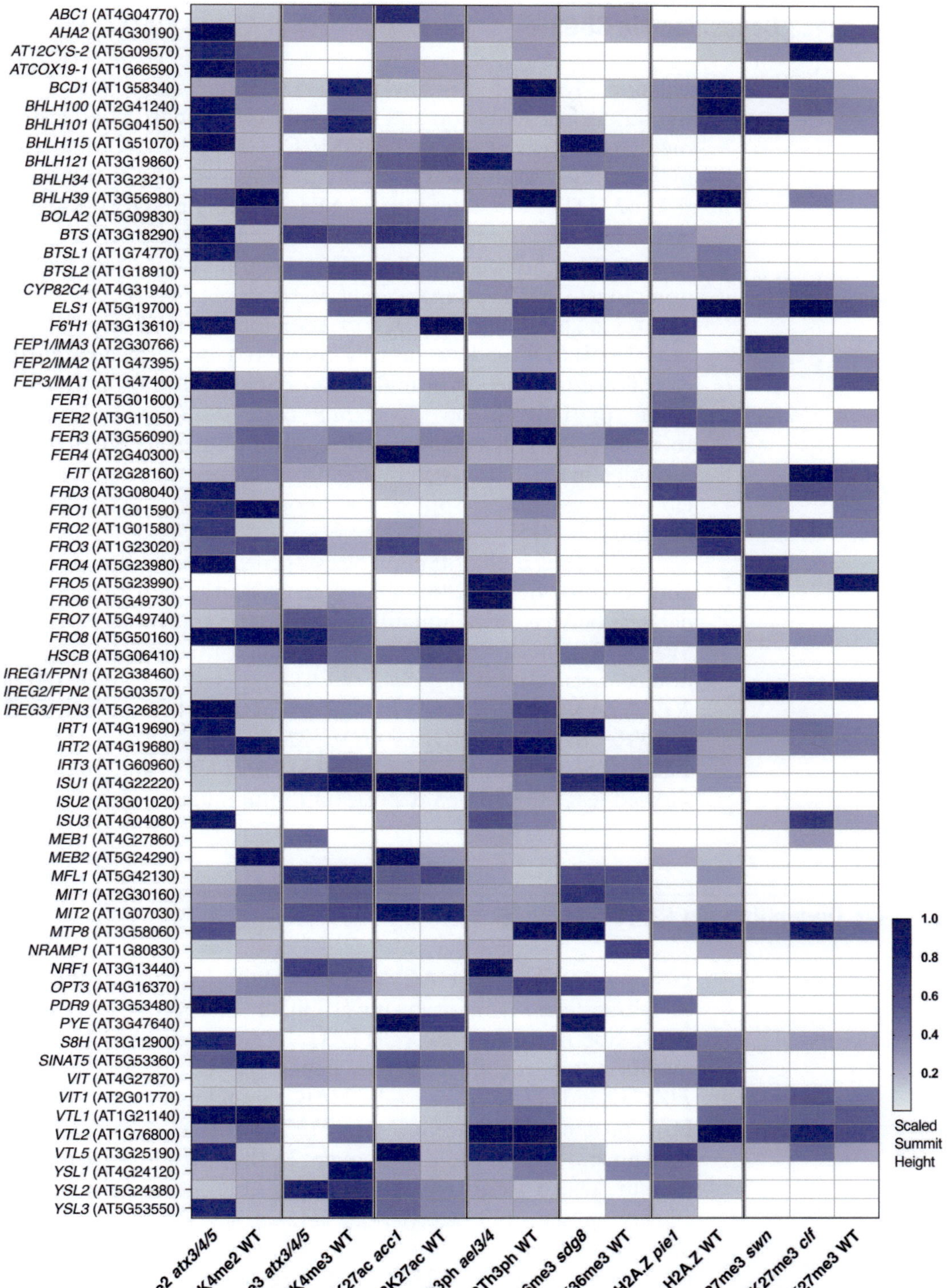

Fig. 2 Heatmap showing the scaled summit height of ChIP-Seq datasets. The summit heights of a total of 66 iron-regulating genes in 8 mutant studies of histone modifying enzymes were collected and transformed into a 0 to 1 range. The darker blue color block represents a more intensive ChIP-seq peak enrichment. Color blocks were set to blank when ChIP-seq signals did not reach the peak-calling threshold.

5 H3K36me3 Deposition via SDG8

Histone 3 lysine 36 trimethylation (H3K36me3) is associated with transcriptional activation and its deposition is mediated by histone methyltransferases [60]. SET DOMAIN GROUP (SDG) is a group of highly conserved methyltransferases in eukaryotes [61, 62]. There are a total of 32 SET-containing SDG proteins in *Arabidopsis thaliana*, which can be grouped into four major classes: Class I, E(z) homologs; Class II, *Ash1* homologs; Class III, *Trx* homologs; Class IV, *Suppressor of variegation* [*Su(var)*] homologs [61, 63, 64]. Among the SDGs in Arabidopsis, SDG8 was classified as a Class II SDG and was initially discovered for its role in inducing early flowering phenotype in the loss of *SDG8*[65, 66] via trimethylation of H3K36 on *Flowering Locus C* (*FLC*) chromatin [67]. SDG8 has been shown to regulate numerous physiological activities in Arabidopsis, ranging from embryonic development, circadian rhythm, response to abiotic stresses, and pathogen defense [68–71].

To investigate if iron homeostasis genes are subjected to epigenetic control by H3K36me3, we analyzed the ChIP-sequencing datasets between *sdg8* mutants and wild type control using dataset SRP149810 from SRA [72]. We identified that loci of nine iron homeostasis-related genes, *FRO7*, *IRT3*, *NATURAL RESISTANCE-ASSOCIATED MACROPHAGE PROTEIN 1* (*NRAMP1*), *OLIGOPEPTIDE TRANSPORTER 3* (OPT3), *YSL1*, *YSL3*, *FERRITIN 3* (FER3), *bHLH34*, and *bHLH115* showed less H3K36me3 deposition in the mutant (Figs. 2 and 3; Table 1).

6 H3ac Deposition via HAC1 and HAC5

Histone acetylation (HAC) proteins belong to the P300/CBP acetyltransferase family, which was initially discovered as an acetyltransferase in mammalian cells [73]. In Arabidopsis, HAC family members are encoded by five genes, *HAC1*, *HAC2*, *HAC4*, *HAC5*, and *HAC12*, encode HAC family members [74]. All HAC proteins share the CBP-type HAT (Histone acetyltransferase) domain and part of the KIX domain. In the Arabidopsis *hac1 hac5* mutant, which lacks two HACs responsible for significant HAT activity, global H3 acetylation (H3ac) level was significantly decreased compared to the level in wild type, indicating their role in H3 acetylation [75]. HAC1, HAC5, and HAC12 have significant HAT activities, while HAC2 does not show any HAT activity [76]. In particular, the *hac1 hac5* double mutant exhibits severe phenotypes, such as smaller and more wrinkly leaves compared to

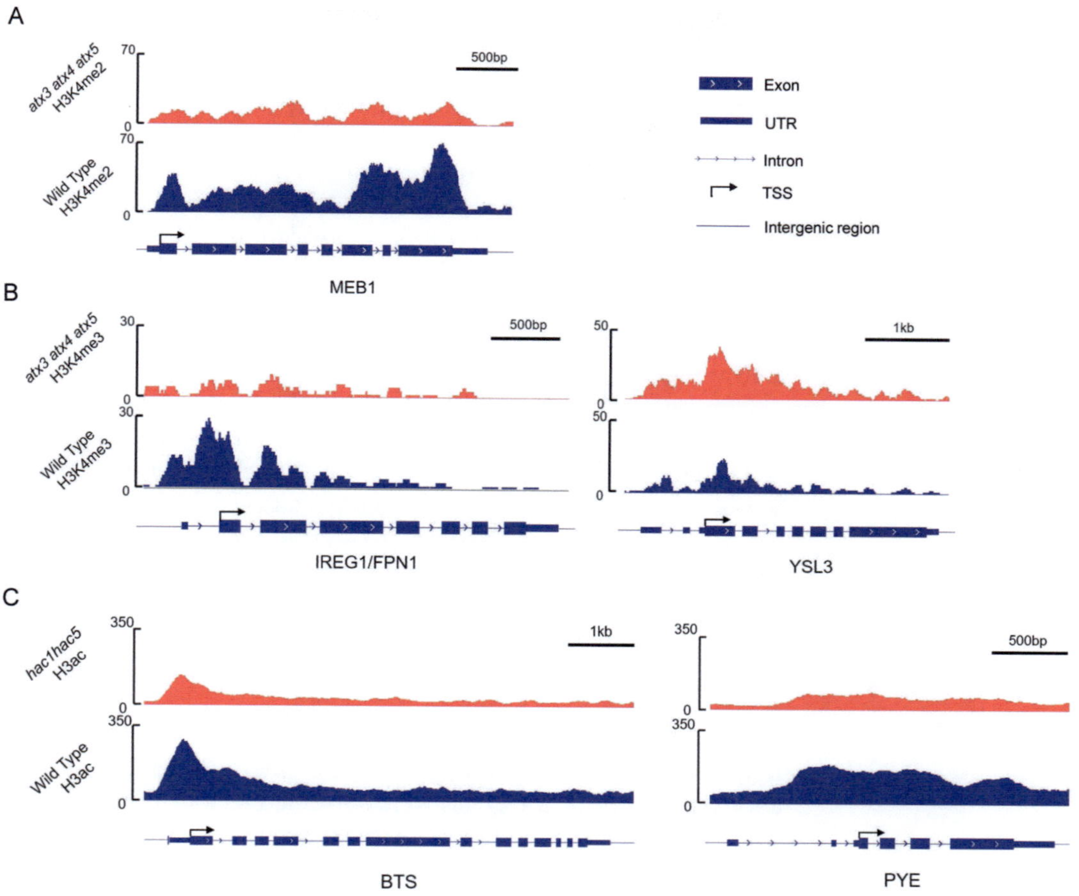

Fig. 3 Genome browser views of histone ChIP peaks (**a**) H3K4me2 in *atx3 atx4 atx5* mutant and wild type, (**b**) H3K4me3 in *atx3 atx4 atx5* mutant and wild type, (**c**) H3Ac in *hac1 hac5* mutant and wild type. The gene tracks are marked at the bottom of the peaks, with introns shown as thinner bands and exons shown as thick bands

hac1 and *hac5* single mutants, suggesting redundant roles of HAC1 and HAC5 [74].

Given the redundancy of HAC1 and HAC5, we compared H3ac depositions detected in wild type and the *hac1 hac5* double mutant using dataset GSE162459 [75]. The comparative analysis of *hac1 hac5* mutant defective in acetyl transferase and wild type revealed differential H3 acetylation status in three iron homeostasis genes, *BCD1*(*BUSH-AND-CHLOROTIC-DWARF 1*), *BTS* (*BRUTUS*), and *PYE* (*POPEYE*) (Table 1, Fig. 3). Notably, PYE is detected as a target of both H3K36me3 and H3ac (Table 1), implying that both of these histone modifications contribute to fine-control the expression of PYE and PYE- dependent iron homeostasis genes.

7 H3Th3ph Deposition via AEL3 and AEL4

Epigenetic regulation by histone phosphorylation affects numerous cellular processes, including signal transduction and mitotic phase transition, and histone phosphorylation is detected in both nucleosome core histones, H3 and H2A, and the linker histone H1 [77–79]. The amino acid targets of phosphorylation are typically serine, threonine, and tyrosine, and their modifications control various transcriptional effects. For instance, H3Th11ph represses the transcription of key genes in root development, whereas H3S10ph activates genes involved in mitosis [80, 81].

Phosphorylation on histone H3 at Th3 can be catalyzed by AtHaspin [79, 81] or Arabidopsis EL-like (AEL) family members [82]. AEL family members have been shown to regulate abiotic stress, vegetative growth, and flowering status [83–85]. In particular, AEL3 and AEL4 are involved in osmotic stress response, and the *ael3 ael4* double mutant exhibited decreased global H3Th3ph level and hypersensitivity under osmotic stress treatment, indicating that these kinases are not only responsible for osmotic stress regulation but might also repress transcription of osmotic stress response genes [83].

Furthermore, among the stress response genes that were differentially expressed in wild type and the *ael3 ael4* double mutant, about one-third of the genes were subjected to antagonistic deposition of H3K4me and H3Th3ph under drought stress [83]. Therefore, we compared the H3Th3ph levels of wild type and *ael3 ael4* using ChIP-Seq dataset GSE68370 [83] to examine if differential deposition of phosphorylation is found on iron homeostasis gene loci. By comparing the phosphorylation status of iron homeostasis-related genes, we noticed that three genes, *IRT2*, *IRT3*, and *IMA1*, were highly phosphorylated at H3Th3 in the wild type, but not in the double mutant. These data imply that AEL3 and AEL4 can be involved in iron homeostasis via controlling H3Th3ph on these loci.

8 H2A.Z Substitution of Nucleosome via PIE1

H2A.Z, a histone variant of H2A associated with nucleosomes near the Transcription Start Site (TSS) is distinguished from H2A by its amino acid sequence and distinct effects on nucleosomes that result in varied gene expression patterns [86]. H2A.Z substitution affects the stability of nucleosomes by the formation of homotypic/heterotypic nucleosomes, regulating rapid transcriptional activation and repression [87]. In Arabidopsis, H2A.Z substitution is catalyzed by the SWR1 complex, which is encoded by the *PIE1* gene [88, 89]. By remodeling chromatin, PIE1 regulates the chromatin's status of various genes, including *FLC* to activate transcription, and plays a vital role in plants' growth and development [90].

We analyzed the H2A.Z abundance difference between wild type and *pie1* mutant based on the ChIP-seq data to monitor potential regulation of iron homeostasis genes using the open database [91]. The results indicate that *PIE1* is involved in the H2A.Z regulation of a vast range of iron homeostasis regulatory genes, including iron uptake, transport, signaling, and transcriptional regulation (Table 1). In Arabidopsis, iron uptake requires three important genes, *AHA2*, *FRO2*, and *IRT1* as introduced previously [92–94], and all three genes, as shown in our results, are potentially regulated by *PIE1* (Fig. 2). This pattern implies that *PIE1* may regulate iron homeostasis through histone variant H2A.Z at the transcriptional level, which might be promoted by other iron uptake-related genes that display differentially enriched HA2.Z in wild type and *pie1* mutant, such as *PDR9* and *IRT2*. The iron homeostasis-related genes identified in our analysis (Table 1) also imply that *PIE1*-related H2A.Z could regulate organellar iron transport, iron storage, or signaling.

9 Discussion

Epigenetic hallmarks on histones convey information that is critical for different types of physiological processes, including responses to diverse abiotic stressors [13, 72, 83]. Here, to reduce the chances of incorporating false positive targets in our outcome, we focused on comparing ChIP-seq data from wild type to data from mutants defective in the direct modifiers of histones, such as methyltransferases, acetyltransferases, or a key component of histone modifying enzyme complex. However, our analysis still has several limitations. First, since the datasets were collected from experiments with whole seedlings, tissue-specific expression of the iron genes cannot be monitored. In addition, the datasets analyzed were obtained from samples grown in regular plant medium with sufficient levels of iron. Thus iron-regulated histone modifications, including transient modifications, would not have been accurately represented in our study. Further studies on tissue-specific histone modifications and gene expression patterns under different iron conditions will be necessary.

Multiple known iron response genes were subjected to more than one type of chromatin modification/remodeling (Fig. 2, Table 1), indicating the potential effects of crosstalk and interactions between different epigenetic modifications on iron homeostasis gene regulation. In conclusion, our survey of ChIP-Seq datasets that focused on the comparative analysis from wild type and chromatin modifier or remodeler mutants in

Arabidopsis suggests multiple candidate genes and potential epigenetic regulators that can be further investigated to better understand crosstalks between epigenetic regulations and iron homeostasis.

Data Availability Statement
The GEO [15] and SRA [16] accession numbers of the ChIP-sequencing datasets used in this study are listed in Table 1.

Author Contributions
Y. Y., Y. W, Z. Y, Z. W., Z. X, and J. L. designed and conducted the analyses, and wrote the manuscript.

Acknowledgments

This study was supported by Duke Kunshan University, 2022 SRS fund, 2021 Interdisciplinary Seed Grant, Synear and Wang-Cai Seed Grant 2022, and Wang-Cai Biochemistry Lab Grant to J. L.

References

1. Cedar H, Bergman Y (2009) Linking DNA methylation and histone modification: patterns and paradigms. Nat Rev Genet 10: 295–304
2. Berger SL (2007) The complex language of chromatin regulation during transcription. Nature 447:407–412
3. Widiez T, El Kafafi S, Girin T, Berr A, Ruffel S, Krouk G et al (2011) High nitrogen insensitive 9 (HNI9)-mediated systemic repression of root NO3- uptake is associated with changes in histone methylation. Proc Natl Acad Sci U S A 108:13329–13334
4. Chandrika NN, Sundaravelpandian K, Schmidt W (2013) A PHD in histone language: on the role of histone methylation in plant responses to phosphate deficiency. Plant Signal Behav 8:e24381
5. Chandrika NNP, Sundaravelpandian K, Yu SM, Schmidt W (2013) ALFIN-LIKE 6 is involved in root hair elongation during phosphate deficiency in Arabidopsis. New Phytol 198:709–720
6. Neto J, da Silva MD, Pandolfi V, Crovella S, Benko-Iseppon AM, Kido EA (2017) Epigenetic signals on plant adaptation: a biotic stress perspective. Curr Protein Pept Sci 18: 352–367
7. Zogli P, Libault M (2017) Plant response to biotic stress: is there a common epigenetic response during plant-pathogenic and symbiotic interactions? Plant Sci 263:89–93
8. Kong L, Liu Y, Wang X, Chang C (2020) Insight into the role of epigenetic processes in abiotic and biotic stress response in wheat and barley. Int J Mol Sci 21
9. Huang CY, Jin H (2021) Coordinated epigenetic regulation in plants: a potent managerial tool to conquer biotic stress. Front Plant Sci 12:795274
10. Secco D, Whelan J, Rouached H, Lister R (2017) Nutrient stress-induced chromatin changes in plants. Curr Opin Plant Biol 39: 1–7
11. Séré D, Martin A (2020) Epigenetic regulation: another layer in plant nutrition. Plant Signal Behav 15:1686236
12. Fan H, Zhang Z, Wang N, Cui Y, Sun H, Liu Y et al (2014) SKB1/PRMT5-mediated histone H4R3 dimethylation of Ib subgroup bHLH genes negatively regulates iron homeostasis in Arabidopsis thaliana. Plant J 77:209–221
13. Park EY, Tsuyuki KM, Parsons EM, Jeong J (2020) PRC2-mediated H3K27me3 modulates shoot iron homeostasis in Arabidopsis thaliana. Plant Signal Behav 15:1784549

14. Su J, Yao Z, Wu Y, Lee J, Jeong J (2022) Minireview: chromatin-based regulation of iron homeostasis in plants. Front Plant Sci 13:959840
15. Barrett T, Wilhite SE, Ledoux P, Evangelista C, Kim IF, Tomashevsky M et al (2013) NCBI GEO: archive for functional genomics data sets—update. Nucleic Acids Res 41:D991–D995
16. Leinonen R, Sugawara H, Shumway M (2011) The sequence read archive. Nucleic Acids Res 39:D19–D21
17. Pickrell JK, Gaffney DJ, Gilad Y, Pritchard JK (2011) False positive peaks in ChIP-seq and other sequencing-based functional assays caused by unannotated high copy number regions. Bioinformatics 27:2144–2146
18. Jain D, Baldi S, Zabel A, Straub T, Becker PB (2015) Active promoters give rise to false positive 'Phantom Peaks' in ChIP-seq experiments. Nucleic Acids Res 43:6959–6968
19. Kouzarides T (2007) Chromatin modifications and their function. Cell 128:693–705
20. Pfluger J, Wagner D (2007) Histone modifications and dynamic regulation of genome accessibility in plants. Curr Opin Plant Biol 10:645–652
21. Eichten SR, Schmitz RJ, Springer NM (2014) Epigenetics: beyond chromatin modifications and complex genetic regulation. Plant Physiol 165:933–947
22. Moronczyk J, Braszewska A, Wojcikowska B, Chwialkowska K, Nowak K, Wojcik AM et al (2022) Insights into the histone acetylation-mediated regulation of the transcription factor genes that control the embryogenic transition in the somatic cells of arabidopsis. Cells 11:11
23. Vaillant I, Paszkowski J (2007) Role of histone and DNA methylation in gene regulation. Curr Opin Plant Biol 10:528–533
24. Cao R, Wang L, Wang H, Xia L, Erdjument-Bromage H, Tempst P et al (2002) Role of histone H3 lysine 27 methylation in Polycomb-group silencing. Science 298: 1039–1043
25. Naumann K, Fischer A, Hofmann I, Krauss V, Phalke S, Irmler K et al (2005) Pivotal role of AtSUVH2 in heterochromatic histone methylation and gene silencing in Arabidopsis. EMBO J 24:1418–1429
26. Zhang K, Sridhar VV, Zhu J, Kapoor A, Zhu JK (2007) Distinctive core histone post-translational modification patterns in Arabidopsis thaliana. PLoS One 2:e1210
27. Xu L, Zhao Z, Dong A, Soubigou-Taconnat-L, Renou JP, Steinmetz A et al (2008) Di- and tri- but not monomethylation on histone H3 lysine 36 marks active transcription of genes involved in flowering time regulation and other processes in Arabidopsis thaliana. Mol Cell Biol 28:1348–1360
28. Padeken J, Methot SP, Gasser SM (2022) Establishment of H3K9-methylated heterochromatin and its functions in tissue differentiation and maintenance. Nat Rev Mol Cell Biol 23:623–640
29. Andrews S (2010) FastQC: A quality control tool for high throughput sequence data. http://www.bioinformatics.babraham.ac.uk/projects/fastqc/
30. Bolger AM, Lohse M, Usadel B (2014) Trimmomatic: a flexible trimmer for Illumina sequence data. Bioinformatics 30:2114–2120
31. Langmead B, Salzberg SL (2012) Fast gapped-read alignment with Bowtie 2. Nat Methods 9:357–359
32. Danecek P, Bonfield JK, Liddle J, Marshall J, Ohan V, Pollard MO et al (2021) Twelve years of SAMtools and BCFtools. Gigascience 10
33. Zhang Y, Liu T, Meyer CA, Eeckhoute J, Johnson DS, Bernstein BE et al (2008) Model-based analysis of ChIP-Seq (MACS). Genome Biol 9:R137
34. Ross-Innes CS, Stark R, Teschendorff AE, Holmes KA, Ali HR, Dunning MJ et al (2012) Differential oestrogen receptor binding is associated with clinical outcome in breast cancer. Nature 481:389–393
35. Wang Q, Li M, Wu T, Zhan L, Li L, Chen M et al (2022) Exploring epigenomic datasets by ChIPseeker. Curr Protocol 2:e585
36. Zhang X, Bernatavichute YV, Cokus S, Pellegrini M, Jacobsen SE (2009) Genome-wide analysis of mono-, di- and trimethylation of histone H3 lysine 4 in Arabidopsis thaliana. Genome Biol 10:R62
37. Du Z, Li H, Wei Q, Zhao X, Wang C, Zhu Q et al (2013) Genome-wide analysis of histone modifications: H3K4me2, H3K4me3, H3K9ac, and H3K27ac in Oryza sativa L. Japonica. Mol Plant 6:1463–1472
38. Alvarez-Venegas R, Pien S, Sadder M, Witmer X, Grossniklaus U, Avramova Z (2003) ATX-1, an Arabidopsis homolog of trithorax, activates flower homeotic genes. Curr Biol 13:627–637
39. Ding Y, Lapko H, Ndamukong I, Xia Y, Al-Abdallat A, Lalithambika S et al (2009) The Arabidopsis chromatin modifier ATX1, the myotubularin-like AtMTM and the response to drought. Plant Signal Behav 4: 1049–1058

40. Liu Y, Zhang A, Yin H, Meng Q, Yu X, Huang S et al (2018) Trithorax-group proteins ARABIDOPSIS TRITHORAX4 (ATX4) and ATX5 function in abscisic acid and dehydration stress responses. New Phytol 217:1582–1597
41. Wang X, Wang D, Xu W, Kong L, Ye X, Zhuang Q et al (2021) Histone methyltransferase ATX1 dynamically regulates fiber secondary cell wall biosynthesis in Arabidopsis inflorescence stem. Nucleic Acids Res 49: 190–205
42. Chen LQ, Luo JH, Cui ZH, Xue M, Wang L, Zhang XY et al (2017) ATX3, ATX4, and ATX5 encode putative H3K4 methyltransferases and are critical for plant development. Plant Physiol 174:1795–1806
43. Oya S, Takahashi M, Takashima K, Kakutani T, Inagaki S (2022) Transcription-coupled and epigenome-encoded mechanisms direct H3K4 methylation. Nat Commun 13:4521
44. Yamada K, Nagano AJ, Nishina M, Hara-Nishimura I, Nishimura M (2013) Identification of two novel endoplasmic reticulum body-specific integral membrane proteins. Plant Physiol 161:108–120
45. Morrissey J, Baxter IR, Lee J, Li L, Lahner B, Grotz N et al (2009) The ferroportin metal efflux proteins function in iron and cobalt homeostasis in Arabidopsis. Plant Cell 21: 3326–3338
46. Kumar RK, Chu HH, Abundis C, Vasques K, Rodriguez DC, Chia JC et al (2017) Iron-nicotianamine transporters are required for proper long distance iron signaling. Plant Physiol 175:1254–1268
47. Wiles ET, Selker EU (2017) H3K27 methylation: a promiscuous repressive chromatin mark. Curr Opin Genet Dev 43:31–37
48. Wang L, Joshi P, Miller EL, Higgins L, Slattery M, Simon JA (2018) A role for monomethylation of histone H3-K27 in gene activity in *Drosophila*. Genetics 208: 1023–1036
49. Shaver S, Casas-Mollano JA, Cerny RL, Cerutti H (2010) Origin of the polycomb repressive complex 2 and gene silencing by an E (z) homolog in the unicellular alga Chlamydomonas. Epigenetics 5:301–312
50. Li C, Chen C, Gao L, Yang S, Nguyen V, Shi X et al (2015) The Arabidopsis SWI2/SNF2 chromatin remodeler BRAHMA regulates polycomb function during vegetative development and directly activates the flowering repressor gene SVP. PLoS Genet 11: e1004944
51. Wang H, Liu C, Cheng J, Liu J, Zhang L, He C et al (2016) Arabidopsis flower and embryo developmental genes are repressed in seedlings by different combinations of polycomb group proteins in association with distinct sets of cis-regulatory elements. PLoS Genet 12: e1005771
52. Chen DH, Huang Y, Jiang C, Si JP (2018) Chromatin-based regulation of plant root development. Front Plant Sci 9:1509
53. Shu J, Chen C, Li C, Cui Y (2020) The complexity of PRC2 catalysts CLF and SWN in plants. Biochem Soc Trans 48:2779–2789
54. Park EY, Tsuyuki KM, Hu F, Lee J, Jeong J (2019) PRC2-mediated H3K27me3 contributes to transcriptional regulation of FIT-dependent iron deficiency response. Front Plant Sci 10:627
55. Chanvivattana Y, Bishopp A, Schubert D, Stock C, Moon YH, Sung ZR et al (2004) Interaction of polycomb-group proteins controlling flowering in arabidopsis. Development 131:5263–5276
56. Wang D, Tyson MD, Jackson SS, Yadegari R (2006) Partially redundant functions of two SET-domain polycomb-group proteins in controlling initiation of seed development in Arabidopsis. Proc Natl Acad Sci U S A 103: 13244–13249
57. Hennig L, Derkacheva M (2009) Diversity of Polycomb group complexes in plants: same rules, different players? Trends Genet 25: 414–423
58. Shu J, Chen C, Thapa RK, Bian S, Nguyen V, Yu K et al (2019) Genome-wide occupancy of histone H3K27 methyltransferases CURLY LEAF and SWINGER in Arabidopsis seedlings. Plant Direct 3:e00100
59. Godwin J, Farrona S (2022) The importance of networking: plant polycomb repressive complex 2 and its interactors. Epigenomes 6
60. Liu C, Lu F, Cui X, Cao X (2010) Histone methylation in higher plants. Annu Rev Plant Biol 61:395–420
61. Springer NM, Napoli CA, Selinger DA, Pandey R, Cone KC, Chandler VL et al (2003) Comparative analysis of SET domain proteins in maize and Arabidopsis reveals multiple duplications preceding the divergence of monocots and dicots. Plant Physiol 132:907–925
62. Qian C, Zhou MM (2006) SET domain protein lysine methyltransferases: structure, specificity and catalysis. Cell Mol Life Sci 63: 2755–2763
63. Baumbusch LO, Thorstensen T, Krauss V, Fischer A, Naumann K, Assalkhou R et al

(2001) The Arabidopsis thaliana genome contains at least 29 active genes encoding SET domain proteins that can be assigned to four evolutionarily conserved classes. Nucleic Acids Res 29:4319–4333

64. Dong H, Liu D, Han T, Zhao Y, Sun J, Lin S et al (2015) Diversification and evolution of the SDG gene family in Brassica rapa after the whole genome triplication. Sci Rep 5:16851
65. Soppe WJ, Bentsink L, Koornneef M (1999) The early-flowering mutant efs is involved in the autonomous promotion pathway of Arabidopsis thaliana. Development 126:4763–4770
66. Cazzonelli CI, Millar T, Finnegan EJ, Pogson BJ (2009) Promoting gene expression in plants by permissive histone lysine methylation. Plant Signal Behav 4:484–488
67. Zhao Z, Yu Y, Meyer D, Wu C, Shen WH (2005) Prevention of early flowering by expression of FLOWERING LOCUS C requires methylation of histone H3 K36. Nat Cell Biol 7:1256–1260
68. Kim SY, Michaels SD (2006) SUPPRESSOR OF FRI 4 encodes a nuclear-localized protein that is required for delayed flowering in winter-annual Arabidopsis. Development 133:4699–4707
69. Grini PE, Thorstensen T, Alm V, Vizcay-Barrena G, Windju SS, Jorstad TS et al (2009) The ASH1 HOMOLOG 2 (ASHH2) histone H3 methyltransferase is required for ovule and anther development in Arabidopsis. PLoS One 4:e7817
70. Berr A, McCallum EJ, Alioua A, Heintz D, Heitz T, Shen WH (2010) Arabidopsis histone methyltransferase SET DOMAIN GROUP8 mediates induction of the jasmonate/ethylene pathway genes in plant defense response to necrotrophic fungi. Plant Physiol 154:1403–1414
71. Yang H, Howard M, Dean C (2014) Antagonistic roles for H3K36me3 and H3K27me3 in the cold-induced epigenetic switch at Arabidopsis FLC. Curr Biol 24:1793–1797
72. Li Y, Brooks M, Yeoh-Wang J, McCoy RM, Rock TM, Pasquino A et al (2020) SDG8-mediated histone methylation and RNA processing function in the response to nitrate signaling. Plant Physiol 182:215–227
73. Ortega E, Rengachari S, Ibrahim Z, Hoghoughi N, Gaucher J, Holehouse AS et al (2018) Transcription factor dimerization activates the p300 acetyltransferase. Nature 562:538–544
74. Li C, Xu J, Li J, Li Q, Yang H (2014) Involvement of arabidopsis HAC family genes in pleiotropic developmental processes. Plant Signal Behav 9:e28173
75. Guo J, Wei L, Chen SS, Cai XW, Su YN, Li L et al (2021) The CBP/p300 histone acetyltransferases function as plant-specific MEDIATOR subunits in Arabidopsis. J Integr Plant Biol 63:755–771
76. Earley KW, Shook MS, Brower-Toland B, Hicks L, Pikaard CS (2007) In vitro specificities of Arabidopsis co-activator histone acetyltransferases: implications for histone hyperacetylation in gene activation. Plant J 52:615–626
77. Brosch G, Georgieva EI, Lopez-Rodas G, Lindner H, Loidl P (1992) Specificity of Zea mays histone deacetylase is regulated by phosphorylation. J Biol Chem 267:20561–20564
78. Zhao J, Grafi G (2000) The high mobility group I/Y protein is hypophosphorylated in endoreduplicating maize endosperm cells and is involved in alleviating histone H1-mediated transcriptional repression. J Biol Chem 275: 27494–27499
79. Ashtiyani RK, Moghaddam AM, Schubert V, Rutten T, Fuchs J, Demidov D et al (2011) AtHaspin phosphorylates histone H3 at threonine 3 during mitosis and contributes to embryonic patterning in Arabidopsis. Plant J 68:443–454
80. Demidov D, Van Damme D, Geelen D, Blattner FR, Houben A (2005) Identification and dynamics of two classes of aurora-like kinases in Arabidopsis and other plants. Plant Cell 17: 836–848
81. Kurihara D, Matsunaga S, Omura T, Higashiyama T, Fukui K (2011) Identification and characterization of plant Haspin kinase as a histone H3 threonine kinase. BMC Plant Biol 11:73
82. Raxwal VK, Ghosh S, Singh S, Katiyar-Agarwal S, Goel S, Jagannath A et al (2020) Abiotic stress-mediated modulation of the chromatin landscape in Arabidopsis thaliana. J Exp Bot 71:5280–5293
83. Wang Z, Casas-Mollano JA, Xu J, Riethoven JJ, Zhang C, Cerutti H (2015) Osmotic stress induces phosphorylation of histone H3 at threonine 3 in pericentromeric regions of Arabidopsis thaliana. Proc Natl Acad Sci U S A 112:8487–8492
84. Kang J, Cui H, Jia S, Liu W, Yu R, Wu Z et al (2020) Arabidopsis thaliana MLK3, a plant-specific casein kinase 1, negatively regulates flowering and phosphorylates histone H3 in vitro. Genes (Basel) 11
85. Wang Z, Kang J, Armando Casas-Mollano J, Dou Y, Jia S, Yang Q et al (2021) MLK4-mediated phosphorylation of histone H3T3 promotes flowering by transcriptional silencing of FLC/MAF in Arabidopsis thaliana. Plant J 105:1400–1412

86. Bargaje R, Alam MP, Patowary A, Sarkar M, Ali T, Gupta S et al (2012) Proximity of H2A. Z containing nucleosome to the transcription start site influences gene expression levels in the mammalian liver and brain. Nucleic Acids Res 40:8965–8978
87. Giaimo BD, Ferrante F, Herchenrother A, Hake SB, Borggrefe T (2019) The histone variant H2A.Z in gene regulation. Epigenetic Chromatin 12:37
88. Luk E, Ranjan A, Fitzgerald PC, Mizuguchi G, Huang Y, Wei D et al (2010) Stepwise histone replacement by SWR1 requires dual activation with histone H2A.Z and canonical nucleosome. Cell 143:725–736
89. Aslam M, Fakher B, Jakada BH, Cao S, Qin Y (2019) SWR1 chromatin remodeling complex: a key transcriptional regulator in plants. Cell 8
90. Noh YS, Amasino RM (2003) PIE1, an ISWI family gene, is required for FLC activation and floral repression in Arabidopsis. Plant Cell 15:1671–1682
91. Luo YX, Hou XM, Zhang CJ, Tan LM, Shao CR, Lin RN et al (2020) A plant-specific SWR1 chromatin-remodeling complex couples histone H2A.Z deposition with nucleosome sliding. EMBO J 39:e102008
92. Vert G, Barberon M, Zelazny E, Seguela M, Briat JF, Curie C (2009) Arabidopsis IRT2 cooperates with the high-affinity iron uptake system to maintain iron homeostasis in root epidermal cells. Planta 229:1171–1179
93. Grillet L, Lan P, Li W, Mokkapati G, Schmidt W (2018) IRON MAN is a ubiquitous family of peptides that control iron transport in plants. Nat Plants 4:953–963
94. Martin-Barranco A, Spielmann J, Dubeaux G, Vert G, Zelazny E (2020) Dynamic control of the high-affinity iron uptake complex in root epidermal cells. Plant Physiol 184:1236–1250
95. Gollhofer J, Timofeev R, Lan P, Schmidt W, Buckhout TJ (2014) Vacuolar-iron-transporter1-like proteins mediate iron homeostasis in Arabidopsis. PLoS One 9:e110468
96. Eroglu S, Meier B, von Wiren N, Peiter E (2016) The vacuolar manganese transporter MTP8 determines tolerance to iron deficiency-induced chlorosis in Arabidopsis. Plant Physiol 170:1030–1045
97. Jain A, Dashner ZS, Connolly EL (2019) Mitochondrial iron transporters (MIT1 and MIT2) are essential for iron homeostasis and embryogenesis in Arabidopsis thaliana. Front Plant Sci 10
98. Green LS, Rogers EE (2004) FRD3 controls iron localization in Arabidopsis. Plant Physiol 136:2523–2531
99. Khan MA, Castro-Guerrero NA, McInturf SA, Nguyen NT, Dame AN, Wang J et al (2018) Changes in iron availability in Arabidopsis are rapidly sensed in the leaf vasculature and impaired sensing leads to opposite transcriptional programs in leaves and roots. Plant Cell Environ 41:2263–2276
100. Khoshnevis S, Dreggors RE, Hoffmann TFR, Ghalei H (2019) A conserved Bcd1 interaction essential for box C/D snoRNP biogenesis. J Biol Chem 294:18360–18371
101. Li Y, Lu CK, Li CY, Lei RH, Pu MN, Zhao JH et al (2021) IRON MAN interacts with BRUTUS to maintain iron homeostasis in Arabidopsis. Proc Natl Acad Sci U S A 118
102. Song Z, Lin S, Fu J, Chen Y, Zhang H, Li J et al (2022) Heterologous expression of ISU1 gene from Fragaria vesca enhances plant tolerance to Fe depletion in Arabidopsis. Plant Physiol Biochem 184:65–74
103. DiDonato RJ Jr, Roberts LA, Sanderson T, Eisley RB, Walker EL (2004) Arabidopsis Yellow Stripe-Like2 (YSL2): a metal-regulated gene encoding a plasma membrane transporter of nicotianamine-metal complexes. Plant J 39:403–414
104. Bauer P, Ling HQ, Guerinot ML (2007) FIT, the FER-like iron deficiency induced transcription factor in Arabidopsis. Plant Physiol Biochem 45:260–261
105. Cai Y, Li Y, Liang G (2021) FIT and bHLH Ib transcription factors modulate iron and copper crosstalk in Arabidopsis. Plant Cell Environ 44:1679–1691
106. Rodriguez-Celma J, Connorton JM, Kruse I, Green RT, Franceschetti M, Chen YT et al (2019) Arabidopsis BRUTUS-LIKE E3 ligases negatively regulate iron uptake by targeting transcription factor FIT for recycling. Proc Natl Acad Sci U S A 116:17584–17591
107. Kim SA, Punshon T, Lanzirotti A, Li L, Alonso JM, Ecker JR et al (2006) Localization of iron in Arabidopsis seed requires the vacuolar membrane transporter VIT1. Science 314:1295–1298
108. Frazzon AP, Ramirez MV, Warek U, Balk J, Frazzon J, Dean DR et al (2007) Functional analysis of Arabidopsis genes involved in mitochondrial iron-sulfur cluster assembly. Plant Mol Biol 64:225–240
109. Siwinska J, Siatkowska K, Olry A, Grosjean J, Hehn A, Bourgaud F et al (2018) Scopoletin 8-hydroxylase: a novel enzyme involved in coumarin biosynthesis and iron-deficiency responses in Arabidopsis. J Exp Bot 69: 1735–1748
110. Tabata R, Kamiya T, Imoto S, Tamura H, Ikuta K, Tabata M et al (2022) Systemic

regulation of iron acquisition by Arabidopsis in environments with heterogeneous iron distributions. Plant Cell Physiol 63:842–854

111. Wang Z, Qian C, Guo X, Liu E, Mao K, Mu C et al (2016) ELS1, a novel MATE transporter related to leaf senescence and iron homeostasis in Arabidopsis thaliana. Biochem Biophys Res Commun 476:319–325

112. Connolly EL, Campbell NH, Grotz N, Prichard CL, Guerinot ML (2003) Overexpression of the FRO2 ferric chelate reductase confers tolerance to growth on low iron and uncovers posttranscriptional control. Plant Physiol 133:1102–1110

113. Jain A, Wilson GT, Connolly EL (2014) The diverse roles of FRO family metalloreductases in iron and copper homeostasis. Front Plant Sci 5:100

114. Gautam CK, Tsai HH, Schmidt W (2021) IRONMAN tunes responses to iron deficiency in concert with environmental pH. Plant Physiol 187:1728–1745

115. Rogers EE, Guerinot ML (2002) FRD3, a member of the multidrug and toxin efflux family, controls iron deficiency responses in Arabidopsis. Plant Cell 14:1787–1799

116. Durrett TP, Gassmann W, Rogers EE (2007) The FRD3-mediated efflux of citrate into the root vasculature is necessary for efficient iron translocation. Plant Physiol 144:197–205

117. Chu HH, Car S, Socha AL, Hindt MN, Punshon T, Guerinot ML (2017) The Arabidopsis MTP8 transporter determines the localization of manganese and iron in seeds. Sci Rep 7:11024

118. Cailliatte R, Schikora A, Briat JF, Mari S, Curie C (2010) High-affinity manganese uptake by the metal transporter NRAMP1 is essential for Arabidopsis growth in low manganese conditions. Plant Cell 22:904–917

119. Chen HM, Wang YM, Yang HL, Zeng QY, Liu YJ (2019) NRAMP1 promotes iron uptake at the late stage of iron deficiency in poplars. Tree Physiol 39:1235–1250

120. Zhai Z, Gayomba SR, Jung HI, Vimalakumari NK, Pineros M, Craft E et al (2014) OPT3 is a phloem-specific iron transporter that is essential for systemic iron signaling and redistribution of iron and cadmium in Arabidopsis. Plant Cell 26:2249–2264

121. Waters BM, Chu HH, Didonato RJ, Roberts LA, Eisley RB, Lahner B et al (2006) Mutations in Arabidopsis yellow stripe-like1 and yellow stripe-like3 reveal their roles in metal ion homeostasis and loading of metal ions in seeds. Plant Physiol 141:1446–1458

122. Ravet K, Touraine B, Boucherez J, Briat JF, Gaymard F, Cellier F (2009) Ferritins control interaction between iron homeostasis and oxidative stress in Arabidopsis. Plant J 57: 400–412

123. Lin YF, Liang HM, Yang SY, Boch A, Clemens S, Chen CC et al (2009) Arabidopsis IRT3 is a zinc-regulated and plasma membrane localized zinc/iron transporter. New Phytol 182:392–404

124. Gao F, Robe K, Gaymard F, Izquierdo E, Dubos C (2019) The transcriptional control of iron homeostasis in plants: a tale of bHLH transcription factors? Front Plant Sci 10:6

125. Seo PJ, Park J, Park MJ, Kim YS, Kim SG, Jung JH et al (2012) A Golgi-localized MATE transporter mediates iron homoeostasis under osmotic stress in Arabidopsis. Biochem J 442:551–561

126. Selote D, Samira R, Matthiadis A, Gillikin JW, Long TA (2015) Iron-binding E3 ligase mediates iron response in plants by targeting basic helix-loop-helix transcription factors. Plant Physiol 167:273–286

127. Hindt MN, Akmakjian GZ, Pivarski KL, Punshon T, Baxter I, Salt DE et al (2017) BRUTUS and its paralogs, BTS LIKE1 and BTS LIKE2, encode important negative regulators of the iron deficiency response in Arabidopsis thaliana. Metallomics 9:876–890

128. Long TA, Tsukagoshi H, Busch W, Lahner B, Salt DE Benfey PN (2010) The bHLH transcription factor POPEYE regulates response to iron deficiency in Arabidopsis roots. Plant Cell 22:2219–2236

129. Tone Y, Kawai-Yamada M, Uchimiya H (2004) Isolation and characterization of Arabidopsis thaliana ISU1 gene. Biochim Biophys Acta 1680:171–175

130. Hirayama T, Lei GJ, Yamaji N, Nakagawa N, Ma JF (2018) The putative peptide gene FEP1 regulates iron deficiency response in Arabidopsis. Plant Cell Physiol 59:1739–1752

131. Li Y, Lu CK, Li CY, Lei RH, Pu MN, Zhao JH et al (2021) IRON MAN interacts with BRUTUS to maintain iron homeostasis in Arabidopsis. Proc Natl Acad Sci U S A 118: e2109063118

132. Rodríguez-Celma J, Chou H, Kobayashi T, Long TA, Balk J (2019) Hemerythrin E3 ubiquitin ligases as negative regulators of iron homeostasis in plants. Front Plant Sci 10:98

133. Guo J, Wei L, Chen S-S, Cai X-W, Su Y-N, Li L et al (2021) The CBP/p300 histone acetyltransferases function as plant-specific MEDIATOR subunits in Arabidopsis. J Integr Plant Biol 63:755–771

Chapter 10

Arabidopsis Micro-grafting to Study the Systemic Signaling of Nutrient Status

En-Jung Hsieh and Louis Grillet

Abstract

Grafting enables the study of systemic signals that plants use to maintain their homeostasis at the level of the whole organism. Several protocols of Arabidopsis grafting have been published over the years. These methods are limited because they either affect the overall behavior of the plant, or their throughput is low. The method presented here is based on grafting 3- to 4-days-old seedlings directly on an agar plate, without the use of hormone or collar, and can produce consistently over a hundred grafted plants per day and operator.

Key words Systemic signaling, Long-distance transport, Whole-plant homeostasis, Phloem, Xylem, Shoot-to-root, Root-to-shoot

1 Introduction

Grafting consists in cutting the aerial part of a plant, called a scion, and placing it onto the rootstock of another plant in order to regenerate a healthy plant bearing the shoots of the former and the roots of the latter. In the field of plant nutrition, reciprocal grafting has been used to demonstrate that the uptake of nutrients by the root is regulated by the shoot demand through long-distance signals circulating through the phloem. Notably, grafting was employed to demonstrate the existence of a shoot-to-root signal controlling iron uptake in *Pisum sativum* [1] and to decipher the complex signaling root-to-shoot and shoot-to-root signaling of the nitrogen status in *Arabidopsis thaliana* [2] and *Medicago truncatula* [3]. The protocol presented here was adapted from the method originally described by Marsch-Martínez et al. [4], compared with other popular methods, and shown to preserve the constitutive systemic signaling of Fe-deficiency response of the *opt3–2* mutant [5]. It was later used by Robe et al. (2021) to

Jeeyon Jeong (ed.), *Plant Iron Homeostasis: Methods and Protocols*, Methods in Molecular Biology, vol. 2665,
https://doi.org/10.1007/978-1-0716-3183-6_10,

demonstrate the xylem mobility of scopolin, a precursor of Fe-binding coumarins [6].

Regenerating a healthy grafted plant from two parts and with a sufficient success rate to measure physiological parameters is challenging. Adding to the difficulty, a specific developmental stage is often required to maintain the consistency of the results with previous experiments. The main issues hampering the success of grafting are the failure of healing at the junction of the two plant parts, and the formation of adventitious roots from the scion that predominate over the rootstock. This can lead to a low overall success rate, causing the technique to become too laborious to be useful, and in a worst-case scenario to false positive results caused by adventitious roots.

Various grafting protocols are readily available. Turnbull et al. (2002) developed a method based on the use of capillaries to stabilize the scion onto the rootstock [7]. The major drawbacks of this method are that it is highly laborious for young seedlings and tends to favor the emergence of adventitious root from the scion at the junction of the hypocotyls. Rus et al. (2006) described a method based on the exogenous application of 6-benzylaminopurine and IAA to promote the healing of the hypocotyl and inhibit the adventitious root development [8]. The major shortcomings of this treatment were that it caused a significant developmental delay in our growth conditions, and after 2 weeks of recovery and transfer to a hormone-free medium, the constitutive Fe deficiency response of the *opt3–2* mutant was suppressed.

In the present protocol, 3- to 4-days-old seedlings grown on agar plates are grafted in a laminar flow hood using a stereomicroscope. The cotyledons are removed, which allows the two half plants to lay completely flat on the agar surface and improves the stability of the graft at the hypocotyl junction without resorting to the use of capillaries. The grafts are assembled on a plate containing a high agar percentage (1.5 to 2%), which facilitates the manipulation of the plantlets by preventing them from sinking into the agar. Sucrose is supplemented at low concentrations to produce healthy plants while preventing them from growing too fast during the healing period. After a 5 days recovery period, the seedlings are transferred back to an optimal growth medium. The grafted plants are usually indistinguishable from ungrafted plants from which the cotyledons were removed.

2 Materials

2.1 Seed Sterilization

1. Sterilization solution: 2% sodium hypochlorite, 70% absolute ethanol. This is prepared by mixing 3 mL of commercial bleach (unscented Clorox, at 5.5–6%) with 7 mL of absolute ethanol.

2. Absolute ethanol. Using 95% ethanol results in residual moisture in the seeds, affecting germination and shortening shelf-life. Ethanol containing traces of methanol is not suitable and will lead to very low germination rates.
3. 1000 and 200 μL pipettes and tips.
4. A beaker to discard liquid waste.
5. Timer.

2.2 Plant Growth Medium

1. Autoclaved wooden toothpicks.
2. Nutrients from Estelle and Somerville (ES) media (9): 2.5 mM KH_2PO_4, 5 mM KNO_3, 2 mM $MgSO_4$, 2 mM $CaNO_3$, 50 μM $Fe^{(III)}$-EDTA, 70 μM H_3BO_3, 14 μM $MnCl_2$, 0.5 μM $CuSO_4$, 1 μM $ZnSO_4$, 0.2 μM $NaMoO_4$, and 0.01 μM $CoCl_2$, buffered at pH 5.5 with 5 mM 2-(N-morpholino)ethanesulfonic acid.

   The stock solutions of macronutrients are prepared at 100× concentration as followed:
   - 250 mM KH_2PO_4
   - 500 mM KNO_3
   - 200 mM $MgSO_4$
   - 200 mM $CaNO_3$

   Two micronutrients stock solutions at 1000× concentrations are prepared as followed:
   - 50 mM $Fe^{(III)}$-Na-EDTA
   - Micronutrient mix: 70 mM H_3BO_3, 14 mM $MnCl_2$, 0.5 mM $CuSO_4$, 1 mM $ZnSO_4$, 0.2 mM $NaMoO_4$, 0.01 mM $CoCl_2$
3. Surgical tape (e.g. 3 M Micropore™) to seal the Petri dishes.
4. 5× square Petri dishes containing ES nutrients, 0.5–1% agar, and moderate sucrose (0.5%) with 15 × 3–5 days old seedlings of genotype 1 grown vertically.
5. 5× square Petri dishes containing ES nutrients, 0.5–1% agar, and moderate sucrose (0.5%) with 15 × 3–5 days old seedlings of genotype 2 grown vertically.
6. 10× grafting plates: square Petri dishes containing ES nutrients, 1.5–2% agar, and 0.5% sucrose.

2.3 Grafting

1. Stereomicroscope, preferably one with a single-arm boom stand to avoid interrupting the airflow.
2. Microscissors (can be replaced by a scalpel but scissors are more convenient)
3. Fine tweezer
4. Scalpel blades No. 11

3 Methods

All the procedures are carried out in aseptic conditions in a horizontal laminar flow hood.

3.1 Seed Sterilization

1. Place a small number of seeds (corresponding to a volume of 20–30 μL) in 1.5 mL microcentrifuge tubes. For a larger quantity of seeds, it is better to use several tubes than a larger tube.
2. Add 1 mL of sterilization solution to each tube and mix gently by inverting the tubes for 5 min.
3. Remove the sterilization solution with a pipette. Pipette carefully to avoid removing seeds.
4. Add 1 mL of ethanol is added into each tube to wash the bleach and mix by inverting for 30 seconds.
5. Repeat the washing 2 more times (3 times in total). The whole process should not last more than 30 min to preserve the germination rate. Tubes are processed in batches of 8 at most.
6. After the last washing step, remove the ethanol and lay the seeds onto the tube wall. The tube is placed vertically on a rack and the residual ethanol accumulating at the bottom of the tube is removed using the 200 μL pipette (Fig. 1).
7. Leave the seeds to dry in the flow hood for at least 2 h and preferably overnight. Seeds are considered dry when they easily detach from the tube wall and fall at the bottom of the tube. Dry sterile seeds maintain a near 100% germination rate for a few months if stored in the dark at 4 °C in a refrigerator.

3.2 Preparing Agar Plates

Plates are best prepared from autoclaved media cooled to 50 °C in a water bath the day before use, and left in the flow hood overnight, covered with plastic to prevent drying. Storing agar plates at 4 °C even for short periods affects the gel texture and causes condensation issues. This is detrimental to graft assembly as well as overall reproducibility.

1. Fill a 1 L beaker with 800 mL of ultrapure water.
2. Add 0.976 g of MES and dissolve it with a magnetic stirrer.
3. Adjust the pH to 5.5 with 3 M KOH.
4. Add 10 mL of each macronutrient stock solution, 1 mL of micronutrient mix, and 1 mL of $Fe^{(III)}$-Na-EDTA.
5. Add 5 g of sucrose and mix until it dissolves. Optimal sucrose concentration might vary depending on the photoperiod, light intensity, and spectrum of the growth chamber.

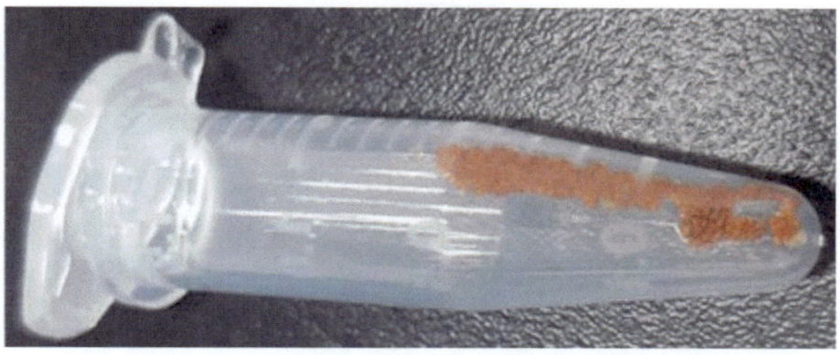

Fig. 1 Ethanol-soaked seeds are laid against the tube wall to allow the ethanol to decant at the bottom of the tube. The tube will be placed vertically on a rack, and the residual ethanol will be removed by pipetting with a thin tip

6. Add 8 g of agar for optimal growth plates and 15 g for grafting plates. Note that the hardness of the gel varies between different batches of agar.
7. Autoclave for 30 min at 115 °C degrees.
8. Place the hot medium in a water bath at 50 °C and wait until the temperature equilibrates.
9. In the flow hood, poor 50 mL of the medium into a sterile 50 mL graduated tube and pour it into the square Petri dish. Poor another 20 mL in the Petri dish. Leave it open for a few minutes until you filled 5 to 10 plates to avoid condensation. Close the plates after a few minutes.

3.3 Plant Growth

1. Sow a line of 15 regularly-spaced seeds of the first genotype on each of the 5 agar plates using the toothpicks humected with the medium.
2. Repeat this step with the seeds of the second genotype and the 5 next agar plates.
3. Place the plates in a growth chamber as vertically as possible to avoid bending the hypocotyls of the seedlings during growth.
4. Let the plants grow for 3 to 5 days. Ideally, the seedlings should have a clearly visible straight hypocotyl and open cotyledons but the leaf primordia should not be visible at the time of grafting. The ideal number of days might vary depending on the light conditions, airflow, and temperature variation within the growth chamber.
5. Place a piece of tape on the label of each Petri dish and wipe the outside of the Petri dish with 70% ethanol before placing them in the flow hood.
6. Cut the cotyledons of all the seedlings in all the plates using micro scissors or scalpel blades. Leave the cotyledons on the plate.
7. On two grafting plates, write the names of the two grafting combinations to be performed, for instance: Wild-Type/

mutant on the first one and mutant/Wild-type on the second one. Prepare a third plate with a line drawn in the middle and the name of each genotype on each side of the plate.

8. With a fine tweezer, carefully carry 5 seedlings (the plants will stick to the tweezer by capillarity) of the first genotype onto the upper part of a plate. Make sure the genotype name is written. Then carry 5 seedlings of the other side of the plate, harboring the corresponding genotype name.
9. Using a scalpel blade, cut all the seedlings in the middle of the hypocotyl with a 90° angle between the blade and the hypocotyl.
10. Transfer 5 rootstocks onto the grafting plate harboring the corresponding rootstock name. Make sure that the *roots are in contact with the medium* to avoid drying.
11. Transfer 5 scions onto this plate at a distance of about 0.5–1 cm over each of the hypocotyls of the rootstocks.
12. At this stage, the 5 rootstocks and 5 scions on the grafting plates are ready to assemble. The plate containing the reciprocal scions and rootstocks (the ones not in use yet) must stay closed at all times and can wait until the first plate is processed.
13. Place the grafting plate under the stereomicroscope and adjust the focus and magnification. A 2 to 3× magnification is enough at this stage.
14. Move a scion gently by pushing it with the tweezer until its cut hypocotyl is in contact with the hypocotyl of the rootstock. Repeat the operation for the 4 other grafts. Use higher magnification to make sure the two parts are indeed in contact and push gently on the top of the scion to establish a firm contact between both sides.
15. Then take the plate with the remaining scions and rootstocks and assemble them in the same way.
16. *Very carefully tape the plates* while keeping them horizontal, and transport them to the growth chamber, and place the plates as vertically as possible. Do not keep plates horizontally under the hood, because phototropism will cause the hypocotyls to bend within 1–2 h, which will cause the two plant parts to disassemble.
17. Place the plates as vertically as possible in the growth chamber, i.e. with a 90° angle between the plate and the shelf, and manipulate the plate very carefully.
18. Repeat **steps** 7 to **17** for each genotype combination to be studied, typically 4 for a reciprocal grafting experiment between two genotypes. Plates containing plants without cotyledons but not grafted can be added as controls, in addition to

self-grafted seedlings, in order to evaluate the stress effect of the grafting process.

19. After 5 to 7 days, transfer the plants to a new plate containing a regular growth medium.

4 Notes

1. The growth stage of the seedlings is critical. Grafting must be performed before the first leaves are visible at all. It still works afterward, but the success rate is decreased and there is an increased risk of root growing from the scion. This is due to growing leaves pushing on the media and causing the separation of the stock and the scion.
2. In the days following germination, you aim at having plants that grow slowly. On one hand, it widens the window to perform the grafting, on the other hand, it improves the recovery rate by slowing down leaf emergence prior to hypocotyl healing.
3. The ideal number of days after germination depends on the growth conditions, notably light intensity, photoperiod, and sucrose concentration. In general, working with 3 to 5 days-old seedlings is optimal, with some variability between growth chambers. On day 3, seedlings might be small and difficult to handle, whereas on day 5, the leaves might have appeared already. Smaller seedlings with no apparent leaves are preferred.
4. The optimal sucrose concentration depends on the photoperiod. Under continuous light, 0.5% sucrose is low (optimal growth is Columbia-0 is achieved with 1.5% sucrose). With a photoperiod of 16 h/8 h, 0.5% sucrose is optimal for growth, and you might consider using lower concentrations.
5. The seedling must be relatively straight. If the hypocotyl is bent, grafts are more difficult to assemble. This is optimized by sowing 15–20 seeds on one line per germination plate, and picking the straightest ones. Plates should be placed almost vertically (5–10° tilt) to avoid the hypocotyl bending due to gravitropism and phototropism. Seed quality and agar content do affect the curving of the hypocotyl.
6. The solidity of the media is important. Germination should occur on a soft media (0.5–1% agar), hypocotyl section, and assembly of grafts is facilitated by a harder media (1.5–2%). Agar is not a homogenous compound and there are large differences in hardness between batches.
7. When assembling the graft, the contact between the scion and the rootstock must be robust. The straighter the cutting, the

easier to assemble. Rootstocks are positioned first, then the scion is gently pushed onto it, and pressed. This is tricky but improves quickly with practice. If one graft is difficult, just give up and take another one. Ideally, the firmness of the contact is verified by using the highest magnification of the stereomicroscope. This magnification is too high for grafting, but at lower magnification, it is easy to miss that the two hypocotyls are not indeed in contact.

8. Healing takes 5 to 7 days. Plants transferred too early can easily break at the hypocotyl junction. It is preferable to wait longer than to rush the transfer.
9. In reciprocal grafting experiments, it is important to include self-grafted plants of each studied genotype. With this particular protocol, it is also useful to add ungrafted control plants that had their cotyledons removed and were subjected to similar growth conditions than grafted plants. This control allows to assess how much the grafting and healing affected the studied physiological process.
10. Verifying the success of the grafts and specifically the absence of adventitious roots using a stereomicroscope is of utmost importance to avoid false positive results. Any graft with adventitious roots should be excluded from the experiment to ensure the integrity of the conclusions.

References

1. Grusak MA, Pezeshgi S (1996) Shoot-to-root signal transmission regulates root Fe(III) reductase activity in the dgl mutant of pea. Plant Physiol 110:329–334. https://doi.org/10.1104/pp.110.1.329
2. Tabata R, Sumida K, Yoshii T et al (2014) Perception of root-derived peptides by shoot LRR-RKs mediates systemic N-demand signaling. Science 346:343–346. https://doi.org/10.1126/science.1257800
3. Gautrat P, Laffont C, Frugier F (2020) Compact root architecture 2 promotes root competence for nodulation through the miR2111 systemic effector. Curr Biol. https://doi.org/10.1016/j.cub.2020.01.084
4. Marsch-Martínez N, Franken J, Gonzalez-Aguilera KL et al (2013) An efficient flat-surface collar-free grafting method for Arabidopsis thaliana seedlings. Plant Methods 9:14. https://doi.org/10.1186/1746-4811-9-14
5. Grillet L, Lan P, Li W et al (2018) IRON MAN is a ubiquitous family of peptides that control iron transport in plants. Nat Plants 4:953–963. https://doi.org/10.1038/s41477-018-0266-y
6. Robe K, Conejero G, Gao F et al (2021) Coumarin accumulation and trafficking in Arabidopsis thaliana: a complex and dynamic process. New Phytol 229:2062–2079. https://doi.org/10.1111/nph.17090
7. Turnbull CGN, Booker JP, Leyser HMO (2002) Micrografting techniques for testing long-distance signalling in Arabidopsis. Plant J 32:255–262. https://doi.org/10.1046/j.1365-313x.2002.01419.x
8. Rus A, Baxter I, Muthukumar B et al (2006) Natural variants of AtHKT1 enhance Na+ accumulation in two wild populations of Arabidopsis. PLoS Genet 2:e210. https://doi.org/10.1371/journal.pgen.0020210

Chapter 11

Advances in Iron Retrograde Signaling Mechanisms and Uptake Regulation in Photosynthetic Organisms

Maria A. Pagani and Diego F. Gomez-Casati

Abstract

Iron (Fe) is an essential metal for the growth and development of different organisms, including plants and algae. This metal participates in different biological processes, among which are cellular respiration and photosynthesis. Fe is found associated with heme groups and as part of inorganic Fe-S groups as cofactors of numerous cellular proteins. Although Fe is abundant in soils, it is often not bioavailable due to soil pH. For this reason, photosynthetic organisms have developed different strategies for the uptake, the sensing of Fe intracellular levels but also different mechanisms that maintain and regulate adequate concentrations of this metal in response to physiological needs. This work focuses on discussing recent advances in the characterization of the mechanisms of Fe homeostasis and Fe retrograde signaling in photosynthetic organisms.

Key words Iron, Retrograde signaling, Iron deficiency

1 Introduction

Iron (Fe) is an essential micronutrient for all organisms, from prokaryotes to eukaryotes, and one of the most abundant elements on earth. This metal has the ability to be oxidized and reduced, as well as to bind to different molecules, forming cofactors such as heme groups, present in different proteins (i.e., cytochromes) or Fe-S groups, present in ferro-sulfoproteins. In photosynthetic organisms, this metal plays an important role in different processes of cell metabolism involved in growth and development, such as photosynthesis, mitochondrial respiration, DNA synthesis and repair, biogenesis of Fe-S proteins and hemoproteins, and lipid metabolism, among others (Fig. 1) [1, 2].

Photosynthetic organisms must finely regulate Fe levels by controlling its uptake and cellular distribution. There are two major problems with Fe as a free ion: its insolubility and its toxicity. In an oxidizing medium, the concentration of soluble Fe is low, and the Fe present in soils forms oxide or hydroxide Fe-precipitates

Jeeyon Jeong (ed.), *Plant Iron Homeostasis: Methods and Protocols*, Methods in Molecular Biology, vol. 2665, https://doi.org/10.1007/978-1-0716-3183-6_11,

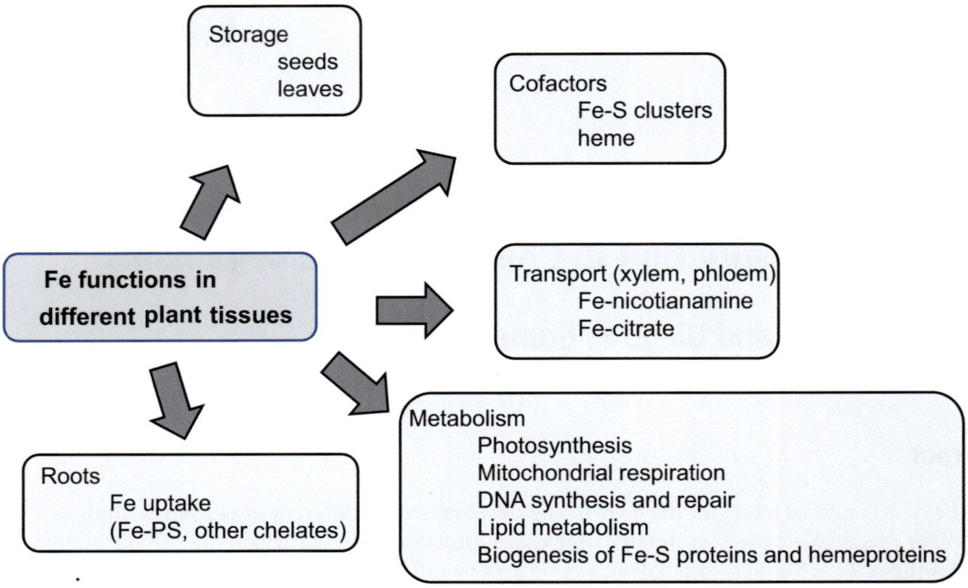

Fig. 1 Participation of Fe in different plant tissues and cellular processes

[3]. Plants have developed two strategies to ensure Fe uptake through the roots. Strategy I, also called the reduction strategy, is carried out by dicots and non-gramineous monocots and strategy II, or chelation strategy, is used by grasses [2, 4, 5]. *Arabidopsis thaliana* is one of the organisms that use Strategy I [4]. *Arabidopsis* plants improve Fe uptake through three reactions: (i) Proton pumping through a protein with ATPase activity (H^+-ATPase2, AHA2). This mechanism promotes the acidification of the medium, thus increasing the solubility and mobilization of Fe^{3+} [6]; (ii) Subsequently, Fe^{3+} is reduced to a more soluble form, Fe^{2+}, by an NADPH-dependent ferric chelate reductase, Ferric Reductase Oxidase 2 (FRO2), present on the cell surface. Electrons are transferred from NADPH through four heme groups, in order to reduce Fe^{3+} [7] and (iii) finally, Fe^{2+} is imported by specific transporters such as IRT1 (Fe-regulated transporter 1). This protein is the most relevant factor in the Fe-uptake process, not being specific to Fe, since it also transports other divalent metals such as Zn^{2+}, Mn^{2+}, Co^{2+}, and Cd^{2+} [8, 9]. It was reported that these three proteins increase their activities under Fe deficiency [10].

The absorption, accumulation, and Fe transport are regulated by the action of different transcription factors, belonging to the bHLH (basic helix-loop-helix) family such as FIT (Fe deficiency-induced transcription factor) [11]. This protein is a member of the bHLH subgroup IIIa+c. FIT participates in the control of Fe uptake by regulating the expression of *IRT1* and *FRO2* [12–14].

On the other hand, in Strategy II, the roots of some plants such as wheat (*Triticum aestivum*) and maize (*Zea mays*) release phytosiderophores which chelate Fe^{3+} in the rhizosphere [5]. Phytosiderophores are a class of molecules belonging to the mugineic acid (MA) family and are synthesized from S-adenosyl methionine (SAM). The enzyme nicotianamine synthase (NAS) catalyzes the combination of three SAM molecules to form nicotianamine (NA), an L-α-amino acid whose function is to chelate metals. NA is a precursor for MA synthesis via the NA aminotransferase enzyme [15, 16]. Thus, phytosiderophores are released from the root epidermis via anion channels or vesicles. After the binding of these compounds to the metal, the Fe-PS complex is translocated into the plant through transporter proteins such as yellow stripe1 (YS1, present in corn) or yellow stripe-like 15 (OsYSL15, present in rice) [17, 18].

It was recently shown that some grasses such as rice, despite using chelation strategy II, also induce the expression of two Fe^{2+} transporters, OsIRT1 and OsIRT2 under Fe deficiency. In this way, it was described that rice plants use a combined strategy to absorb Fe. This may be due to the fact that rice is adapted to grow in conditions of water excess (submerged) where Fe^{2+} would be more abundant than Fe^{3+}. To date, *O. sativa* is the unique plant that uses the combined Fe uptake strategy [19].

2 Systemic Plant Response to Fe Deficiency

As mentioned above, due to the essential role that Fe plays in plant cells, its concentrations in the different compartments must be maintained at an optimal level for its proper functioning. The mechanism of Fe incorporation in plants must reach a balance between its uptake under conditions of metal deficiency and stopping its incorporation under Fe excess. This has resulted in a complex mechanism for regulating Fe metabolism. in plants [20].

The incorporation of Fe in plants is regulated by transcriptional and post-transcriptional mechanisms that cause an increase in the incorporation of the metal in deficiency conditions [21]. Experimental evidence suggests the existence of a shoot-to-root regulatory mechanism that would depend on the availability of Fe in the leaves for their metabolic processes [22]. However, the signals involved in the shoot-to-root regulation process have not yet been clearly elucidated. Probably, this regulation affects the production and/or translocation of activating or repressive signals sent from the aerial part of the plants to the root [1].

Two models have been postulated for systemic signal regulation. The inductive model and the repressive model. The inductive model is based on the fact that the signal is produced in the leaves in response to Fe deficiency and translocated through the phloem to the roots, resulting in the activation of several Fe uptake genes. In this scheme, a high content of Fe in the leaves could prevent the release of the signal and these genes could not be expressed. The opposite occurs in the repressive model, in which the root Fe uptake genes are expressed constitutively, except when the Fe content in the leaves is high. In this case, if Fe levels in the leaves are sufficient, the signal could be activated and negatively regulate the transcription of the root Fe uptake genes. In contrast, the absence of this signal would occur under Fe deficiency and could downregulate gene expression. However, it has not been elucidated so far which of the two models is functional in plants [23].

3 Regulation of the Response to Fe Deficiency

In Strategy I plants, the key regulator of Fe uptake is the transcription factor FIT. In *A. thaliana*, FIT is expressed in the differentiation zone of the root epidermis and is induced under Fe deficiency conditions, and, as mentioned above, positively regulates many genes involved in Fe uptake [11–13]. To fulfill this function, FIT is activated by forming heterodimers through the interaction with transcription factors of the Ib bHLH subgroup such as AtbHLH38, -39, -100 and -101 [24–26]. Using transcriptomic studies, it has been shown that FIT would control the expression of more than 400 genes [12, 27]. Among the genes whose expression is controlled by FIT are the transcription factors MYB10 and MYB72. Both proteins are functionally redundant and are essential for plant growth and survival in alkaline soils with low Fe availability [28]. MYB10 and MYB72 regulate the expression of NA synthase isoform 4 (NAS4). It was shown that *myb10/myb72* double mutants are deficient in *NAS4* expression and both *myb10/myb72* and *nas4-1* mutant plants show low Fe and chlorophyll concentrations, as well as impaired growth [28]. Later on, it was shown that MYB10 and MYB72, together with FIT, control the biosynthesis and excretion of coumarin compounds. These metabolites from the phenylpropanoid pathway are essential to mobilize Fe from recalcitrant pools and would increase Fe solubility by chelation and/or reduction (reviewed by [29]). Under normal Fe conditions, FIT would be inactive and would be ubiquitinated and degraded via the 26S proteasome complex [12, 13, 30].

On the other hand, it was described that *Arabidopsis* plants deficient in FIT expression (*fit1* mutants), showed a decrease in *FRO2* transcripts and ferric chelate reductase activity, leading to the accumulation of low Fe levels in leaves and seeds and also showing severe chlorosis. In contrast, *IRT1* expression was not affected in *fit1* plants; however, IRT1 protein was not detected [12, 13]. These results indicate that *FRO2* is directly regulated by FIT1, while a FIT-dependent mechanism would control the post-transcriptional regulation of *IRT1* [8, 12]. It has been observed that *Arabidopsis* plants that overexpress *FRO2* only show an increase in its expression levels when these plants are subjected to Fe deficiency; indicating that *FRO2* could also be regulated by a post-transcriptional mechanism [8].

As mentioned above, there is another group of genes that modulate the acquisition of Fe through the inactivation or recycling of FIT. Under Fe-deficient conditions, about twenty E3-ubiquitin ligases are induced, including BRUTUS (BTS) in *Arabidopsis* and HRZ1 and HRZ2 (hemerythrin motif-containing RING- and zinc (Zn)-finger proteins) in rice [12, 31, 32]. It was described that the BTS and HRZ proteins negatively regulate Fe accumulation and homeostasis since *bts* and *hrz* mutant lines accumulated Fe and exhibited increased tolerance to Fe deficiency [32]. It was also shown in *Arabidopsis* that BTS proteins associate with some bHLH-type transcription factors such as bHLH105 and -115 which positively regulate Fe signaling [33].

Recently, the presence of two new E3-ubiquitin ligases in *A. thaliana* called BRUTUS-Like1 and -2 (BTSL1 and BTSL2, respectively) was described [32]. These proteins are co-expressed together with other genes that regulate Fe uptake, and also interact with FIT, promoting its degradation. It was also described that although the *btsl1/btsl2* mutant failed to down-regulate the FIT-controlled genes, leading to the accumulation of toxic levels of Fe in roots and leaves, it is more tolerant to Fe deficiency [32].

The mechanism of activation of some transcription factors such as bHLH101 and bHLH115 is currently unclear. Grillet et al. recently identified a family of small Fe uptake-inducing peptides that control metal transport in plants named IRON MAN (IMA) [34, 35]. In *Arabidopsis thaliana*, eight members of the IMA family (IMA1–8) were identified. It was shown that the overexpression of IMA peptides increases the expression of *IRT1* and *FRO2* genes, promoting Fe uptake in plants. Furthermore, these peptides act by promoting the transcription of other *bHLH Ib* genes, indicating that they act upstream of this family of transcription factors [33, 35–39].

Recently, Li et al. showed that IMA peptides as well as bHLH105 and bHLH115 proteins contain a particular BTS-interacting domain (called the BID domain) [35]. Both bHLH factors mentioned and at least six of the IMA peptides (except IMA5 and IMA8), are degraded by BTS [35]. Under Fe-deficient conditions, an increase in the expression of the transcripts that code for IMAs was observed. Thus, they interfere with the interaction between BTS and the bHLH105 and -115 proteins. In this way, the IMAs would act as inhibitors of BTS, reducing the degradation of bHLH105 and -115 and thus, activating the response to Fe deficiency [35]. Although the mechanism of action of IMA peptides and how these peptides regulate Fe homeostasis in plants remains to be clarified, it was proposed that the interactions between these transcription factors are essential to maintain Fe homeostasis under conditions of low Fe content.

There are other bHLH-type factors that are involved in the signaling in response to Fe deficiency. Using transcriptomic techniques, Long et al. [38] identified a new bHLH-type transcription factor called POPEYE (PYE), belonging to the bHLH IVb subgroup. This factor would be also important in the positive regulation of plant growth and development under Fe deficiency. Likewise, PYE would regulate the expression of a set of Fe homeostasis genes related to metal redistribution rather than uptake, as well as other genes involved in the stress response, through interaction with other bHLH factors such as ILR3 (IAA-Leu Resistant 3 protein), which also interacts with BTS [38]. Among the PYE-regulated genes are *NAS4* and *FRO3*, a mitochondrial Fe reductase located in the root vasculature [38, 40, 41].

Recently, a bHLH transcription factor, Upstream Regulator of IRT1 (URI/bHLH121), was identified by two independent studies [42, 43]. While there are some contradictory results, there is consensus that URI/bHLH121 plays a key role in the Fe deficiency signaling cascade in *A. thaliana* (reviewed in [44]). During Fe deficiency, a phosphorylated form of URI forms heterodimers with subgroup IVc bHLH transcription factors. These heterodimers induce the expression of subgroup Ib bHLH genes. Then, subgroup Ib transcription factors (like bHLH38, 38, 100, and 101) and FIT heterodimers induce the expression of Fe uptake genes, such as *IRT1* and *FRO2*.

A good question is how transcription factors recognize certain target sequences to activate transcription. It was shown by chromatin precipitation and immunodetection experiments that bHLH104 and -115 factors, as well as PYE and FIT proteins, interact with the promoter regions of their target genes by binding to E-box or G-box regions [38, 45–48]. It was also shown that FIT stability is regulated by the interaction with other transcription factors such as those of the EIL family, involved in ethylene signaling (EIN3, ethylene insensitive 3 and EIL1, ethylene insensitive 3 like 1) as well as ZAT12, a C2H2-like transcription factor that contains a Zn-finger domain [49, 50]. FIT interaction with EIN3 and/or EIL1 also promotes Fe acquisition during the early stage of Fe deficiency [49].

In conclusion, numerous transcription factors involved in Fe uptake and accumulation in plants have been identified and characterized. However, how the expression of the factors that act first, at the top of the regulatory cascade, is regulated remains to be clarified. Knowing in more detail the mechanism of the activation cascade and the function of the different transcription factors will give us a better understanding of how plants regulate Fe homeostasis.

4 Iron and Retrograde Signaling

Plants contain two organelles that generate energy, chloroplasts and mitochondria, which are also involved in the synthesis of a variety of essential metabolites, including Fe-S clusters, heme, and chlorophyll. Although these organelles contain their own genomes, most of their proteins are encoded in the nucleus. In order to adapt the organelles functioning to the cellular demands, or the developmental or environmental conditions, they need to communicate with the nucleus. Retrograde signaling refers to the linear communication of chloroplasts and mitochondria with the nucleus, leading to the synthesis of proteins required by the organelles. In the last two decades, many plant retrograde pathways have been characterized, mostly involving signaling from the chloroplasts [51, 52]. Additionally, the original concept of retrograde signaling has been broadened to include all responses, from transcription to posttranslational modifications that ultimately impact the organelles' functions, as a result of a stimulus perceived in them and communicated through one or more signals [53].

The electron transport chains from chloroplasts and mitochondria are abundant in all three Fe cofactors (heme, nonheme, and Fe-S clusters), therefore these organelles are the main sites of cellular accumulation of the metal [54]. In order to perform photosynthesis and oxidative phosphorylation, mitochondria and chloroplasts need an adequate supply of the nutrient. When Fe is scarce, photosynthesis is altered resulting in chlorosis (yellowing of leaves) and huge agronomic losses, while mitochondrial respiration is less affected [55, 56]. Moreover, an increase or decrease in the expression of the chloroplast PIC1 Fe permease significantly alters chloroplast development and plant growth through reactions triggered by Fe toxicity or Fe deficiency [57]. Loss of function mutants of MIT, the mitochondrial Fe transporter for Fe uptake, show lower chlorophyll content and altered ferritin gene expression because they accumulate high shoot levels of Fe [58]. Considering the high Fe demand of chloroplasts and mitochondria, and the impact of an inadequate supply of the nutrient in their functioning, it is conceivable that Fe homeostasis mechanisms might involve signals originating in both organelles [59]. A balanced intracellular Fe status requires the ability to sense Fe, but may also rely on indirect signals that report on the physiological processes connected to Fe homeostasis. Two of the most studied and well-characterized plant retrograde signaling pathways are closely related to Fe metabolism. These are the PAP-SAL and the heme/tetrapyrroles retrograde signaling pathways. Furthermore, chloroplastic and mitochondrial Fe-S cluster biogenesis or a signal derived from the pathways also impacts plant Fe homeostasis and the expression of nuclear-encoded chloroplastic and mitochondrial genes [60], in a way that resembles or has elements of retrograde signaling.

5 PAP-SAL1 Pathway

Each retrograde signaling pathway can be considered as made up of different steps or actions that occur sequentially. These steps include (i) signal initiation in organelles, (ii) signal movement or transduction, (iii) perception in the cytosol and/or transduction to the nucleus, and (iv) signal perception leading to transcriptional or posttranscriptional responses [53]. Many of the components or mechanisms involved in each step are currently unknown for the majority of the retrograde signaling pathways described. However, in the PAP-SAL1 pathway, the sequence elements are almost complete.

PAP (3′-phosphoadenosine 5-phosphate) is a by-product of sulfur assimilation reactions, produced in the cytosol from the metabolite PAPS (3′-phosphoadenosine 5′-phosphosulfate). PAP can diffuse to mitochondria and chloroplasts via two bidirectional transporters PAPST1 and PAPST2. PAPST1 exports PAPS from chloroplasts, which exchanges it mainly with ATP or to a lesser extent with PAP. But the most active PAP transporter is the isoform PAPST2, localized to both chloroplasts and mitochondria. PAPST2 imports PAP to the organelles in exchange for ADP, ATP, and PAPS (only in chloroplasts) [61]. PAP is degraded to AMP and phosphate in chloroplasts and mitochondria by the dual localized phosphatase SAL1 (salt tolerance in yeast, 3′(2′),5′--bisphosphate nucleotidase), preventing its accumulation in the cytosol [62]. SAL1 can accept various metabolites for hydrolysis, including PAP, PAPS, and inositol 1,4,5-triphosphate, thus the enzyme is likely involved in several metabolic processes in plant cells, including sulfur (S) metabolism [62, 63]. SAL1 has a homolog in yeast, HAL2, which when inhibited by lithium, accumulates PAP [64]. Similarly, when chloroplastic SAL1 activity is inhibited, PAP accumulates in the cytosol and moves to the nucleus. PAP inhibits the action of 5′-3′ exoribonucleases (XRNs), both nuclear XRN2/XRN3 and cytosolic XRN4, which ultimately results, among other effects, in the expression of plastid redox-associated nuclear genes (PRANGs) [62].

SAL1 is the key to this retrograde signaling pathway acting as an oxidative stress sensor in plant chloroplasts. It senses changes in the photosynthetic redox status, and hydrogen peroxide and superoxide concentrations via redox regulatory mechanisms, which reduce its activity by conformational changes induced by dimerization, intramolecular disulfide formation, and glutathionylation [65]. Under such circumstances PAP accumulates and exerts its inhibitory action over XRNs, altering multiples processes associated with RNA metabolism like posttranscriptional gene silencing (XRN2 and XRN3), rRNA processing (XRN2), and mRNA turnover in stress bodies (XRN4) [66–68]. Hence, the mobile retrograde signal PAP activates the expression of the PRANGs and oxidative stress-responsive genes at least in part by repressing XRNs [62].

In this retrograde signaling pathway, the signal is PAP; however, it is not its synthesis but its degradation that is regulated in the chloroplast (and probably mitochondria). The sensor is the same enzyme that operates on the extinction of the signal, the phosphatase SAL1. Several thiol residues of the enzyme are oxidized in situations of oxidative stress, causing a decrease in its activity, and consequently an increase in the signal. The signal transporters have been identified (PAPSTL2 and to a lesser extent PAPSTL1), although in this case the direction of transport is reversed (from the cytosol to the organelles) since the signal originates passively and/or independently of the stress in the cytosol. The signal impacts nuclear enzymes, the exoribonucleases XRN2 and XRN3, although it may also affect the cytosolic exoribonuclease XRN4, inhibiting their action. The least known step of the pathway is the mechanism of action of these enzymes; however, the inhibition of their functions causes, among other effects, the activation of PRANGs, such as APX2, ELP2, and ZAT10 [62]. The pathway is complex because retrograde signaling involves the decrease or inhibition of the degradation of an inhibitory signal, but as mentioned above, the elements that make up the pathway are almost fully understood. In addition, there is probably no doubt that the dual localized enzyme SAL1 and the PAP transporter PAPST2 must be part of a retrograde signaling pathway. A global transcriptomic metadata study revealed that the PAP-SAL pathway is possibly more related to mitochondrial than chloroplast perturbations [69]. However, there is little information about the role of mitochondria in this pathway, and the stress signals sensed in the organelle.

It was reported that PAP accumulates during high light and drought stress, and *sal1* and *xrn Arabidopsis* mutants, which have constitutive PAP accumulation and/or reduced XRN activity, exhibit increased resistance to such stresses [62]. Likewise, micronutrient availability, and particularly Fe, is one important environmental variable that alters mitochondrial and chloroplast function, with concomitant cell oxidative stress [62], and *sal1*, *xrn2/xrn3*, and xrn4 *Arabidopsis* mutants showed increased resistance to Fe deficiency [70, 71]. Mutant plants show constitutive activation of the root Fe uptake system, with increases in *IRT1*, *FRO2*, and *FIT* transcripts. In addition, an increase in FRO2 activity in roots was also detected, which is considered the rate-limiting step in Fe acquisition [72]. These effects are verified under conditions of deficiency or sufficiency of Fe in the medium, and lead to the accumulation of the metal in the roots, leaves, and seeds of the mutant plants. However, the potential toxic effects of increased Fe in different tissues are mitigated by the increased expression of genes encoding the Fe storage/detoxification proteins FER1 and FER4, and plants show no symptoms of damage by excess Fe.

Another scenario in which the bioavailability of Fe is low is when the growth medium has an alkaline pH. Under these conditions, Fe absorption mechanisms complementary to Strategy I are activated, since the latter is insufficient to absorb the metal. As mentioned above, in *Arabidopsis*, fluorescent coumarin compounds, derived from the phenylpropanoid pathway, are secreted into the medium to solubilize Fe, either by chelation and/or reduction of Fe^{3+} to Fe^{2+}. Recent results showed that mutant plants in the PAP-SAL pathway have also this Fe uptake mechanism activated [73]. The genes for the synthesis of coumarin compounds are induced, as well as the transcription factors that control them. Accumulations of fluorescent compounds were observed in mutant plant roots, and their growth medium showed increased fluorescence, which resulted in greater absorption of Fe. As a consequence, mutant plants in the PAP-SAL pathway better resist alkaline growth conditions, giving more vigorous plants with higher chlorophyll content. It has been suggested that the increase in compounds of the phenylpropanoid pathway may be mediated by small-RNAs (sRNA), since mutants defective in sRNA gene silencing, such as *ago1*, *rdr6* and *tho*, are unable to increase the synthesis of phenolic compounds in certain stress situations [74]. Genes that participate in the PAP/SAL1 retrograde signaling pathway have been identified as suppressors of sRNA-mediated gene silencing [66]. Therefore, mutants in this pathway have increased sRNA-mediated gene silencing and might explain why an increased accumulation of phenolics under alkaline conditions was observed. Ultimately, the effects observed are the result of the PAP accumulation and the inhibition of XRNs- activities that, in wild-type lines, respond to changes in the activity of the sensor enzyme SAL1.

It was also reported that mutant plants in the PAP/SAL pathway absorbed Fe at a higher rate than wild-type plants, even when Fe levels became elevated. Repressive Fe signals or systems, that normally would be expected to limit excessive Fe uptake, were switched off or ineffective. There is little information about the shutdown of these signals once the plant recovers the adequate nutritional Fe status, but PAP levels might be one of the signals sensed. Low levels of PAP might reduce the Fe uptake activity. It is known that Fe scarcity produces extensive metabolic remodeling in chloroplasts and mitochondria, and as a constituent of the electron transport chains, its deficiency causes changes in cellular redox status, resulting in reactive oxygen species (ROS) accumulation [75]. The SAL1 enzyme, as a sensor of the redox state of the chloroplast, decreases its activity under oxidative stress and allows the accumulation of PAP [65]. Once Fe levels bounce back, it is expected that oxidative stress decreases, and, in wild-type plants, the SAL1 activity increases, leading to a decrease in PAP levels. It will be very important to complete the validation of this Fe sensing and signaling mechanism to determine SAL1 activity and PAP levels in wild-type plants subjected to different Fe nutritional situations.

PAP-SAL signaling does not seem to be involved in the root-shoot regulatory mechanism of Fe homeostasis [70]. Foliar Fe application to mutant plants grown under Fe deficiency was able to decrease the expression of the genes that code for FIT and the bHLH transcription factors 38 and 39, which interact with the first, inducing the transcription of the Fe uptake genes, among others. Therefore, the aerial part of the plants was capable of sensing the Fe that was added, and sending the still unknown signal through the phloem. In the roots, the signal could also be integrated, since there was a response at the transcriptional level. However, Fe uptake could not be deactivated, since FRO2 activity remained elevated, as were *FRO2* and *IRT1* transcript levels. There are a few possible explanations for this outcome. Low amounts of active FIT may be enough to activate *FRO2* and *IRT1*. Alternatively, the transcription of *FRO2* and *FIT* could be activated by a different set of transcription factors, partially or totally independent of FIT.

There is also evidence that the alteration in Fe homeostasis in PAP-SAL mutants is related to the action of ethylene. The addition of an ethylene synthesis inhibitor inactivated the induction of Fe uptake genes *IRT1* and *FRO2* and decreased FRO2 activity in all mutants [70]. XRN4 is also known as EIN5 (ethylene insensitive 5) [76], and is involved in the degradation of the transcription factor ERF1 (ethylene-responsive factor 1), which is part of the ethylene signaling [77]. In the PAP-SAL1 mutants, there is an increase in *ERF1* transcripts, which could be one of the factors that activate the transcription of *IRT1* and *FRO2*, since both have at least 9 ERF boxes in their promoters. In addition, protein extracts from wild-type plants, but not those from *xrn4* plants, can decrease the levels of *ERF1* transcripts in RNA decay assays. Interactions between PAP accumulating mutants and jasmonic (JA) and abscisic acids (ABA) have been reported [78, 79]. Thus, the PAP-SAL1 pathway was proposed as a case study for integrating chloroplast retrograde signaling and hormonal signaling in plant growth and morphogenesis [80]. A consistent increase in ABA, gibberellin, and auxin has been reported for *sal1* mutants, whereas ethylene levels were not assessed [80]. Auxin, ethylene, and nitric oxide (NO) act as activators of the Fe deficiency responses, and there is a complex interplay among them, each one activating the synthesis and/or other steps of their signaling cascades, which results in the amplification of the signals [81]. It would be interesting to analyze ethylene levels in

PAP-SAL1 mutants, which could be the link with Fe metabolism, although the increased auxin levels in these mutants also provide evidence of the possible connection with Fe homeostasis regulation.

6 Tetrapyrroles and Retrograde Signaling

Retrograde signaling mechanisms have been classified as biogenic – those related to de novo biogenesis, fundamentally of chloroplasts – and operational – those related to the functional adaptation of organelles and cell metabolism to developmental stages or environmental conditions. The tetrapyrrole pathway is fundamentally associated with biogenic mechanisms, although it is also involved in operational signaling. This is one of the most studied signaling pathways, and yet there are still controversies regarding the participation and/or role of some metabolites. Much of what is known about this pathway comes from studying plants treated with the herbicide norfluorazan [82]. This compound inhibits the synthesis of carotenoids, producing photobleaching and inhibiting the synthesis of photosynthesis-associated nuclear genes (PhANGs) such as those from the chlorophyll-binding LHCb gene family [83]. By using the herbicide, it was possible to identify 6 GUN (genome uncoupled) mutants that have derepressed expression of PhANGs when chloroplasts are damaged, and 5 of them are related to the tetrapyrrole pathway [83–85].

In plants, the pathways of heme, chlorophyll, and phytochromobilins synthesis are localized in plastids, and all branch off from the plastid tetrapyrrole pathway [86–88]. The pathway begins with three enzymatic steps whereby glutamate is used to form aminolevulinic acid (ALA), the tetrapyrrole precursor [87]. Eight molecules of ALA are used to form uroporphyrinogen III, which has the basic tetrapyrrole-conjugated ring structure. The pathway branches at uroporphyrinogen III to form siroheme, or protoporphyrin IX (PPIX), the common precursor for chlorophyll and heme production [87]. Fe insertion into PPIX by ferrochelatase (FC) leads to heme formation while Mg^{2+} insertion and three or four enzymatic steps lead to functional chlorophylls a and b, respectively [89]. Phytochromobilins are synthesized from heme by two enzymatic reactions. Notably, several enzymes of heme and chlorophyll metabolism are Fe-S-cluster-dependent enzymes, linking the metabolisms of both Fe cofactors. Furthermore, some enzymes occur in several isoforms, possibly with differential expression in development or in different tissues, and some enzymes, especially those involved in the heme pathway, have dual localization in chloroplasts and mitochondria. Protoporphyrinogen IX oxidase (PPO) has two isoenzymes with a dual location [90], and FC also has two isoenzymes, although there is controversy about its dual location (see below) [91].

Gun mutants demonstrated the relationship between tetrapyrrole metabolism and retrograde signaling. GUN4 and GUN5 participate in the synthesis of Mg-PPIX [92, 93], and GUN2, GUN3, and GUN6 belong to the heme pathway [85]. GUN2 (heme oxygenase) and GUN3 (phytochromobilin synthase) are responsible for the degradation/transformation of heme into the phytochrome chromophore phytobilin, and *gun6* is a mutant with increased expression of ferrochelatase 1 (FC1). There has been great controversy regarding the identity of the tetrapyrrole signal metabolite, with some early studies proposing Mg-PPXI [94]; however, several reports suggest a lack of correlation between this metabolite level and nuclear gene expression [86, 95, 96]. Currently, the most accepted model is that a specific heme pool generated by flux through FC1 functions as a positive signal to promote the expression of genes required for chloroplast development. In addition to the positive heme signal operating under normal developmental conditions, negative signals may be produced by aberrant chloroplast development accumulating chlorophyll intermediates like Mg-PPIX. Photo-excitation of free porphyrins generates ROS, and these signals could be rapidly transduced to downregulate nuclear gene expression. Thus, the tetrapyrrole pathway may provide both positive and inhibitory signals to control the expression of nuclear genes [84].

On the other hand, a complex but interesting model for the study of the general alteration of metabolism, especially Fe metabolism, and retrograde signaling from both mitochondria and chloroplasts is that of plants with decreased expression of frataxin (*atfh*). The deletion of the gene is lethal at the level of the embryos, which gives indications of the essential functions that the protein fulfills in *Arabidopsis* plants. *atfh* plants showed slight increases in Fe content in mitochondria, chloroplasts and at the whole plant level, and in ferritins 1 and 4 [60, 97], but there was also an increase in the number of root hairs and the level of NO [97]. This shows that there was an alteration in Fe metabolism because responses associated with both deficiency and excess of the metal were seen. Chlorophyll content was practically normal, with a slight decrease in advanced stages of development, and the CO_2 fixation rate was higher than that of wild plants. However, there was retarded growth, reduced fresh weight of fruits, and reduced number of seeds per fruit [98, 99], providing hints that the mitochondrial function was more affected than that of the chloroplast. *Atfh* plants showed decreased activities of Fe-S enzymes from both mitochondria (e.g., aconitase and succinate dehydrogenase), and chloroplast -nitrite reductase and ferredoxin [98]. The decrease in the activity of chloroplast enzymes led to verifying the subcellular localization of frataxin, and it was confirmed that it has a dual localization in

Arabidopsis [60]. This double localization in chloroplasts and mitochondria was also confirmed in maize, which has two frataxin isoforms and both are located in the two organelles [100]. In addition, the nitrite reductase activity of protein extracts from *atfh1* plant chloroplasts could be recovered when incubated with recombinant frataxin preloaded with Fe [60], providing evidence of its functions and location.

Studies in yeast lacking frataxin showed that mitochondrial Fe is unavailable for heme synthesis, suggesting that frataxin could have a role as a mitochondrial Fe donor involved in heme metabolism [101]. Human frataxin interacts with FC [102], and knocking down the expression of frataxin revealed a reduction of heme a and in the cytochrome oxidase activity, suggesting an important role of frataxin in the biogenesis of heme-containing proteins [103]. *atfh* plants showed markedly decreased levels of heme, protoporphyrin pathway enzyme transcripts, and catalase activity (CAT). CAT activity could be recovered when protein extracts were incubated with hemin [104], confirming that the deficiency was in the cofactor biosynthesis. In plants, two ferrochelatases (FC1 and FC2) catalyze the incorporation of Fe into PPIX in the last step of heme synthesis in chloroplasts. Despite that FC are absent from other cell compartments, about 30% of cellular ferrochelatase activity has been observed in plant mitochondria [105, 106]. Recombinant AtFH was able to catalyze the formation of heme in vitro when it was incubated with Fe and PPIX. When frataxin was combined with AtNFS1 (ISC cysteine desulfurase) and AtISD11 (ISC NFS stabilizer protein), the FC activity was increased [107]. These results suggest that frataxin could be the Fe donor in the final step of heme synthesis in plant mitochondria, and possibly the protein responsible for the FC activity measured in mitochondria.

Frataxin is a ubiquitous protein that is present in most organisms, from bacteria and fungi to mammals and plants. The structure of frataxin has been conserved throughout evolution, suggesting that it could have the same function in all organisms. Several roles have been assigned to this protein including participation in Fe homeostasis, Fe–S cluster assembly and biosynthesis, respiration and oxidative phosphorylation, regulation of respiration and control of antioxidant defenses and Fe chaperone, among others, but its exact function has remained elusive and highly debated since its discovery [108]. Currently, it is widely accepted that frataxin functions primarily in Fe-S cluster biosynthesis, increasing cysteine desulfurase activity [109]. However, and without discarding the consensus function, in the *Arabidopsis* decreased expression model there is evidence for a potential role in the heme pathway, as a Fe donor. If AtFH is responsible for the FC activity in mitochondria, and perhaps also in chloroplasts, low levels of this protein would lead to an accumulation of PPIX. This could alter the flux in the tetrapyrrole pathway, shifting the equilibrium towards the formation of MgPPIX or other intermediates. MgPPIX may be an inhibitory signal per se, or through photooxidation and ROS generation, of the nuclear gene expression. Fe accumulation might be the consequence of futile cycles of deficiency signals associated with intermediate metabolites of the tetrapyrrole cycle-, increased uptake, and inability to synthesize the final product, heme.

Another model to study Fe retrograde signaling through metabolites from the tetrapyrrole pathway is the circadian clock. It was postulated that chloroplasts are the central components for circadian sensing of Fe levels in photosynthetic organisms [110]. This organelle contains about 80% of the total Fe present in the leaves and has the function, among others, of storing this metal but also being the site of incorporation of Fe to other molecules such as chlorophylls, heme, and Fe-S centers.

Using some circadian clock reporter genes such as CCA1 (circadian clock associated 1), LHY (late elongated hypocotyl), PPR5 -7 -9 (pseudo-response regulator 5, -7 or -9) and TOC1 (timing of CAB 1). It was described that Fe levels affect the period of the circadian rhythm in *Arabidopsis.* Under limiting Fe conditions, a lengthening of the circadian phase was observed, while the circadian period was shorter under normal or excess Fe conditions [110, 111]. It was also shown that when combined Fe deficiency and impaired chloroplast function did not show additive effects on the circadian period, suggesting that there would be a retrograde chloroplast-nucleus signaling pathway that would be involved in the shortening of the circadian period under Fe deficiency. The nature of this possible signal remains to be elucidated, based on these results it was postulated that there would be a sensor that would be essential to couple the circadian clock with Fe levels and that it would strongly depend on the functional and developmental state of the chloroplasts and the presence of phytochromes [110–112].

The reason why plants modulate the circadian period under Fe deficiency could be an adaptive strategy to continue accumulating this metal while remaining in the vegetative phase and thus, reduce the possibility of producing immature fruits and seeds [110–112].

7 Fe-S Cluster Biogenesis and Retrograde Signaling

Fe-S clusters are essential and versatile cofactors, participating in numerous cellular reactions such as electron transfer, enzyme catalysis, and environmental sensing, and are among the oldest cofactors used during the evolution of organisms. Their synthesis can occur spontaneously in vitro from mineral salts and a reducing agent [113], but since they show great lability towards oxygen, and free iron (Fe^{2+}) and sulfides (S^{2-}) are highly toxic, there are complex protein systems dedicated to this function. Fe-S cluster biogenesis is a conserved pathway inherited from the original endosymbiont organisms that gave rise to modern chloroplasts and mitochondria. In plant cells, there is one complete system in the chloroplast, the SUF (sulfur assimilation) pathway, another complete system in mitochondria, the ISC (Fe—S cluster) pathway, and one incomplete system in the cytosol, the CIA (cytosolic Fe–S cluster assembly), which depends on the ISC system for the export of a yet unknown S compound, and it is responsible for the biogenesis of nuclear and cytosolic Fe-S clusters [114]. The ISC and SUF machineries share a similar assembly process. The first step is the sulfur acquisition from L-cysteine via a cysteine desulfurase and Fe from a still unknown Fe donor. The second step is the assembly of the Fe–S cluster on a scaffold protein (in ISC the first cluster formed is the 2Fe-2S and in the SUF system probably it is 4Fe-4S). The third and fourth steps are the Fe–S cluster release and transfer to the apo-target or onto a carrier protein where the preassembled Fe-S cluster can be oxidized or reduced to different forms, as needed, and then transferred to apo-proteins [114]. Most of the proteins that are part of these pathways are essential, as is the synthesis of Fe-S clusters in each subcellular compartment. The most obscure and still controversial reaction is the source of Fe. The candidate protein that accumulates the most evidence as a Fe donor is frataxin [108, 115, 116].

As explained above, the mechanisms of Fe sensing in plant cells are unknown. They have been better characterized in other eukaryotic organisms, and many are related to Fe-S clusters, taking advantage of their sensing properties and their close relationship with Fe metabolism and/or the essential cellular functions that the cofactor performs. In *Saccharomyces cerevisiae,* the mitochondrial synthesis of Fe-S clusters controls both the activation of the Fe deficiency regulon and the response to high Fe [117]. The Fe scarcity sensing mechanism involves the functionality of the ISC pathway in mitochondria and the CIA Fe-S synthesis in cytosol, from which GRX3/GRX4 glutarredoxins obtain their cofactor and the ability to bind AFT1/AFT2 transcription factors. In this way, transcription factors are retained in the cytosol, preventing the activation of the low Fe response [118]. Likewise, in mammals, the synthesis of Fe-S clusters also controls most aspects of Fe metabolism through the bifunctional cytosolic enzyme ACO1/IRP1 [119]. When bound to the 4Fe-4S cluster, the enzyme functions as aconitase, but loss of the cluster leads to structural changes and affinity for IRE (Fe-responsive elements) in the 5′ and 3′ regions of mRNAs related to Fe metabolism, where it controls stability and translation. Neither of these Fe regulatory mechanisms is conserved in plants [120], nonetheless, in light of the available evidence, the possibility of a Fe–S cluster-dependent Fe-sensing mechanism in plants cannot be ruled out. Fungi and fruit flies that fail to biosynthesize Fe-S clusters exhibit abnormalities in intracellular Fe distribution and misregulated activation of high-affinity cellular Fe uptake [121], and similar phenotypes were observed in several *Arabidopsis* lines with altered expression of ISC and SUF pathway genes.

HSCB is the essential mitochondrial cochaperone involved in the transfer of Fe-S to apo-proteins. The gain and loss of function *HSCB* lines showed altered distribution of Fe between roots and shoots [122]. *HSCB* overexpressing plants accumulated Fe in the roots, while the knockdown lines had increased Fe in the shoots compared to wild-type lines. Both *HSCB* and *hscb* lines had stunted growth and lower chlorophyll levels. Different experiments, including foliar Fe application and Fe deficiency treatment, indicated that the shoot-directed control of Fe uptake in roots functions properly in these lines, implying that Fe–S clusters are not involved in this regulatory mechanism. However, the activity of mitochondrial Fe-S enzymes in different tissues correlated with the Fe content, and not with the *HSCB* transcript levels. It was proposed that there is a root-to-shoot homeostatic control of Fe translocation, possibly dependent on the activity of HSCB, which has a conserved rubredoxin domain that might bind Fe or Fe-S [123] and/or Fe-S levels [122].

NSF1, the ISC cysteine desulfurase, is another protein studied in overexpressing and knockdown *Arabidopsis* lines [124, 125]. Besides changes in *ISC* gene expression, which accompanied the *NFS1* transcript levels, these plants showed alterations not only in Fe contents but also in S levels. Overexpressing lines accumulated increased amounts of Fe and S and the mutant plant had lower contents of S. Fe and S uptakes, assimilation, and regulation genes were up-regulated in *NFS1* plants and down-regulated in *nfs1* plants. Close relationships between Fe and S metabolisms have also been reported in different plant species [126–128]. It has been suggested that the deficiency of Fe and S might be responsible for retrograde signaling pathways from the mitochondria to the nucleus. This idea is based on that both Fe and S deficiencies affect mitochondrial respiration [129], and that the Fe and S demands for Fe–S cluster biosynthesis constitute a feedback signal that coordinates the uptake and reduction of both nutrients [126]. Results from *Arabidopsis* altered expression *NFS1* lines [107] confirm that ISC might play a central role in the co-regulation of Fe and S through retrograde signaling, as has been hypothesized [58, 59, 130].

SUFB, SUFC, and SUFD form a protein complex with a stoichiometry of 1:2:1 and function as the scaffold for Fe-S cluster assembly in SUF system [131]. Knockout lines are lethal; RNAi inducible lines in each gene showed highly reduced chlorophyll levels only in the developing leaves -indicating that *SUFB*, *SUFD*, and *SUFC* expression is essential during de novo chlorophyll synthesis and chloroplast biogenesis [132], and higher chlorophyll a/b ratios compared to WT, often observed under conditions in which chlorophyll synthesis is impaired [133]. Levels of each SUFB, SUFD, and SUFC protein were decreased in all RNAi-silenced leaves, indicating an interdependence for the accumulation of SUF proteins, as was observed for ISC genes in *NFS1* lines [124]. Furthermore, the proteins levels for all major types of Fe-S proteins, including PsaA/B, PsaC subunits of PSI, ferredoxin, and the Rieske subunit of the cytochrome b6f complex, were severely decreased in all RNAi plants, while those for LHC proteins showed moderate decrease [132]. It is evident that blocking the SUF pathway impacts all aspects of chloroplast development, from chlorophyll biosynthesis to PS complexes and stroma protein levels. Moreover, in other studies on *Arabidopsis* wild-type plants grown hydroponically under Fe deficiency, a rapid decrease in SUFA and SUFB proteins and downregulation of *SUFB* transcripts was found [55]. This effect was interpreted as a protection mechanism to prevent the risk of forming incomplete clusters which might generate oxidative damage. There was also a rapid decrease in photosynthesis, affecting mainly ferredoxin and the cytochrome-b_6f complex, and biomass production, which points to a locally sensed chloroplast Fe level and signal transduction to adapt nuclear expression.

Another extremely interesting model of plants with alterations in the metabolism of the SUF pathway, and possibly also in the ISC pathway, which severely impacts the general metabolism of Fe and the organization of chloroplasts is that of the expression of a dominant negative NEET protein [134]. In *Arabidopsis*, the only NEET-type protein has a dual localization in both mitochondria and chloroplasts, and participates in the specific transfer of 2Fe-2S clusters, thanks to the lability with which it binds the cluster through three Cys and one His [135]. In plant lines expressing a NEET protein with a His to Cys mutation, the transfer is blocked by the stabilization of cluster binding [136]. Disrupting the function of AtNEET triggers elevated Fe content in the chloroplasts and at whole plant level, ROS accumulation, chlorosis, structural damage to chloroplasts, and leaf-associated Fe- and Fe-S-deficiency responses, affecting all three systems (SUF, ISC, and CIA) [134]. NEET proteins localize to the outer mitochondrial membranes [137], probably to the outer chloroplast membrane as well, and can transfer its 2Fe-2S cluster to DRE2, a key CIA cytosolic Fe-S assembly factor [134]. It was suggested that Fe might be sensed tightly linked to 2Fe-2S cluster metabolism, and/or the abundance of a particular 2Fe-2S protein(s), and that the Fe deficiency response could be mediated from the chloroplast via a retrograde signal associated with the transfer of this 2Fe-2S cluster type [134].

In summary, in recent years numerous efforts have been made to know in more detail how plants sense Fe levels, how the transcripts of genes involved in Fe uptake are induced, and how this process is regulated. Knowing better the mechanisms of signaling and regulation of Fe homeostasis will allow us to design different strategies to produce plants that grow and develop better under Fe-deficient conditions. On the other hand, based on this knowledge, plant improvement and crop biofortification programs could be implemented. Due to the essential role of Fe in the function of living organisms, these strategies will also result in an improvement in human nutrition and health.

Acknowledgments

MAP and DFGC are research members from CONICET. This research was funded by Agencia Nacional de Promoción Científica y Tecnológica (ANPCyT, PICT 2019 067/PICT 2019 090) and Consejo Nacional de Investigaciones Científicas y Técnicas (PIP 2021-2172).

References

1. Schmidt W (2003) Iron solutions: acquisition strategies and signaling pathways in plants. Trends Plant Sci 8:188–193
2. Connorton JM, Balk J, Rodriguez-Celma J (2017) Iron homeostasis in plants – a brief overview. Metallomics 9:813–823
3. Buckhout TJ, Yang TJ, Schmidt W (2009) Early iron-deficiency-induced transcriptional changes in Arabidopsis roots as revealed by microarray analyses. BMC Genomics 10:147
4. Kobayashi T, Nishizawa NK (2012) Iron uptake, translocation, and regulation in higher plants. Annu Rev Plant Biol 63:131–152
5. Romheld V, Marschner H (1986) Evidence for a specific uptake system for iron phytosiderophores in roots of grasses. Plant Physiol 80:175–180
6. Santi S, Schmidt W (2009) Dissecting iron deficiency-induced proton extrusion in Arabidopsis roots. New Phytol 183:1072–1084
7. Robinson NJ, Procter CM, Connolly EL, Guerinot ML (1999) A ferric-chelate reductase for iron uptake from soils. Nature 397: 694–697
8. Connolly EL, Fett JP, Guerinot ML (2002) Expression of the IRT1 metal transporter is controlled by metals at the levels of transcript and protein accumulation. Plant Cell 14: 1347–1357
9. Vert G, Grotz N, Dedaldechamp F, Gaymard F, Guerinot ML, Briat JF, Curie C (2002) IRT1, an Arabidopsis transporter essential for iron uptake from the soil and for plant growth. Plant Cell 14:1223–1233
10. Hell R, Stephan UW (2003) Iron uptake, trafficking and homeostasis in plants. Planta 216:541–551
11. Bauer P, Ling HQ, Guerinot ML (2007) FIT, the FER-like iron deficiency induced transcription factor in Arabidopsis. Plant Physiol Biochem 45:260–261
12. Colangelo EP, Guerinot ML (2004) The essential basic helix-loop-helix protein FIT1 is required for the iron deficiency response. Plant Cell 16:3400–3412
13. Jakoby M, Wang HY, Reidt W, Weisshaar B, Bauer P (2004) FRU (BHLH029) is required for induction of iron mobilization genes in Arabidopsis thaliana. FEBS Lett 577:528–534
14. Schwarz B, Bauer P (2020) FIT, a regulatory hub for iron deficiency and stress signaling in roots, and FIT-dependent and -independent gene signatures. J Exp Bot 71:1694–1705
15. Herbik A, Koch G, Mock HP, Dushkov D, Czihal A, Thielmann J, Stephan UW, Baumlein H (1999) Isolation, characterization and cDNA cloning of nicotianamine synthase from barley. A key enzyme for iron homeostasis in plants. Eur J Biochem 265:231–239
16. Shojima S, Nishizawa NK, Fushiya S, Nozoe S, Irifune T, Mori S (1990) Biosynthesis of phytosiderophores: in vitro biosynthesis of 2'-deoxymugineic acid from l-methionine

and nicotianamine. Plant Physiol 93:1497–1503

17. Curie C, Panaviene Z, Loulergue C, Dellaporta SL, Briat JF, Walker EL (2001) Maize yellow stripe1 encodes a membrane protein directly involved in Fe(III) uptake. Nature 409:346–349
18. Inoue H, Kobayashi T, Nozoye T, Takahashi M, Kakei Y, Suzuki K, Nakazono M, Nakanishi H, Mori S, Nishizawa NK (2009) Rice OsYSL15 is an iron-regulated iron(III)-deoxymugineic acid transporter expressed in the roots and is essential for iron uptake in early growth of the seedlings. J Biol Chem 284:3470–3479
19. Wairich A, de Oliveira BHN, Arend EB, Duarte GL, Ponte LR, Sperotto RA, Ricachenevsky FK, Fett JP (2019) The combined strategy for iron uptake is not exclusive to domesticated rice (Oryza sativa). Sci Rep 9:16144
20. Darbani B, Briat JF, Holm PB, Husted S, Noeparvar S, Borg S (2013) Dissecting plant iron homeostasis under short and long-term iron fluctuations. Biotechnol Adv 31:1292–1307
21. Garcia-Mina JM, Bacaicoa E, Fuentes M, Casanova E (2013) Fine regulation of leaf iron use efficiency and iron root uptake under limited iron bioavailability. Plant Sci 198:39–45
22. Vert GA, Briat JF, Curie C (2003) Dual regulation of the Arabidopsis high-affinity root iron uptake system by local and long-distance signals. Plant Physiol 132:796–804
23. Takahashi Y, Tokumoto U (2002) A third bacterial system for the assembly of iron-sulfur clusters with homologs in archaea and plastids. J Biol Chem 277:28380–28383
24. Sivitz AB, Hermand V, Curie C, Vert G (2012) Arabidopsis bHLH100 and bHLH101 control iron homeostasis via a FIT-independent pathway. PLoS One 7:e44843
25. Wang N, Cui Y, Liu Y, Fan H, Du J, Huang Z, Yuan Y, Wu H, Ling HQ (2013) Requirement and functional redundancy of Ib subgroup bHLH proteins for iron deficiency responses and uptake in Arabidopsis thaliana. Mol Plant 6:503–513
26. Yuan Y, Wu H, Wang N, Li J, Zhao W, Du J, Wang D, Ling HQ (2008) FIT interacts with AtbHLH38 and AtbHLH39 in regulating iron uptake gene expression for iron homeostasis in Arabidopsis. Cell Res 18:385–397
27. Mai HJ, Pateyron S, Bauer P (2016) Iron homeostasis in Arabidopsis thaliana: transcriptomic analyses reveal novel FIT-regulated genes, iron deficiency marker genes and functional gene networks. BMC Plant Biol 16:211
28. Palmer CM, Hindt MN, Schmidt H, Clemens S, Guerinot ML (2013) MYB10 and MYB72 are required for growth under iron-limiting conditions. PLoS Genet 9:e1003953
29. Tsai HH, Schmidt W (2017) Mobilization of iron by plant-borne Coumarins. Trends Plant Sci 22:538–548
30. Cui Y, Chen CL, Cui M, Zhou WJ, Wu HL, Ling HQ (2018) Four IVa bHLH transcription factors are novel interactors of FIT and mediate JA inhibition of iron uptake in Arabidopsis. Mol Plant 11:1166–1183
31. Kobayashi T, Nagasaka S, Senoura T, Itai RN, Nakanishi H, Nishizawa NK (2013) Iron-binding haemerythrin RING ubiquitin ligases regulate plant iron responses and accumulation. Nat Commun 4:2792
32. Rodriguez-Celma J, Connorton JM, Kruse I, Green RT, Franceschetti M, Chen YT, Cui Y, Ling HQ, Yeh KC, Balk J (2019) Arabidopsis BRUTUS-LIKE E3 ligases negatively regulate iron uptake by targeting transcription factor FIT for recycling. Proc Natl Acad Sci U S A 116:17584–17591
33. Selote D, Samira R, Matthiadis A, Gillikin JW, Long TA (2015) Iron-binding E3 ligase mediates iron response in plants by targeting basic helix-loop-helix transcription factors. Plant Physiol 167:273–286
34. Grillet L, Lan P, Li W, Mokkapati G, Schmidt W (2018) IRON MAN is a ubiquitous family of peptides that control iron transport in plants. Nat Plants 4:953–963
35. Li Y, Lu CK, Li CY, Lei RH, Pu MN, Zhao JH, Peng F, Ping HQ, Wang D, Liang G (2021) IRON MAN interacts with BRUTUS to maintain iron homeostasis in Arabidopsis. Proc Natl Acad Sci U S A 118
36. Li X, Zhang H, Ai Q, Liang G, Yu D (2016) Two bHLH transcription factors, bHLH34 and bHLH104, regulate iron homeostasis in Arabidopsis thaliana. Plant Physiol 170:2478–2493
37. Ling HQ, Bauer P, Bereczky Z, Keller B, Ganal M (2002) The tomato fer gene encoding a bHLH protein controls iron-uptake responses in roots. Proc Natl Acad Sci U S A 99:13938–13943
38. Long TA, Tsukagoshi H, Busch W, Lahner B, Salt DE, Benfey PN (2010) The bHLH transcription factor POPEYE regulates response

to iron deficiency in Arabidopsis roots. Plant Cell 22:2219–2236

39. Zhang J, Liu B, Li M, Feng D, Jin H, Wang P, Liu J, Xiong F, Wang J, Wang HB (2015) The bHLH transcription factor bHLH104 interacts with IAA-LEUCINE RESISTANT3 and modulates iron homeostasis in Arabidopsis. Plant Cell 27:787–805
40. Jeong J, Connolly EL (2009) Iron uptake mechanisms in plants: functions of the FRO family of ferric reductases. Plant Sci 176:709–714
41. Klatte M, Schuler M, Wirtz M, Fink-Straube-C, Hell R, Bauer P (2009) The analysis of Arabidopsis nicotianamine synthase mutants reveals functions for nicotianamine in seed iron loading and iron deficiency responses. Plant Physiol 150:257–271
42. Gao F, Robe K, Bettembourg M, Navarro N, Rofidal V, Santoni V, Gaymard F, Vignols F, Roschzttardtz H, Izquierdo E, Dubos C (2020) The transcription factor bHLH121 interacts with bHLH105 (ILR3) and its closest homologs to regulate iron homeostasis in Arabidopsis. Plant Cell 32:508–524
43. Lei R, Li Y, Cai Y, Li C, Pu M, Lu C, Yang Y, Liang G (2020) bHLH121 functions as a direct link that facilitates the activation of FIT by bHLH IVc transcription factors for maintaining Fe homeostasis in Arabidopsis. Mol Plant 13:634–649
44. Riaz N, Guerinot ML (2021) All together now: regulation of the iron deficiency response. J Exp Bot 72:2045–2055
45. Gao F, Robe K, Gaymard F, Izquierdo E, Dubos C (2019) The transcriptional control of iron homeostasis in plants: A tale of bHLH transcription factors? Front Plant Sci 10:6
46. Liang G, Zhang H, Li X, Ai Q, Yu D (2017) bHLH transcription factor bHLH115 regulates iron homeostasis in Arabidopsis thaliana. J Exp Bot 68:1743–1755
47. Yang Y, Ou B, Zhang J, Si W, Gu H, Qin G, Qu LJ (2014) The Arabidopsis Mediator subunit MED16 regulates iron homeostasis by associating with EIN3/EIL1 through subunit MED25. Plant J 77:838–851
48. Zhang Y, Wu H, Wang N, Fan H, Chen C, Cui Y, Liu H, Ling HQ (2014) Mediator subunit 16 functions in the regulation of iron uptake gene expression in Arabidopsis. New Phytol 203:770–783
49. Lingam S, Mohrbacher J, Brumbarova T, Potuschak T, Fink-Straube C, Blondet E, Genschik P, Bauer P (2011) Interaction between the bHLH transcription factor FIT and ethylene insensitive3/ethylene insensitive3-like1 reveals molecular linkage between the regulation of iron acquisition and ethylene signaling in Arabidopsis. Plant Cell 23:1815–1829
50. Liu W, Karemera NJU, Wu T, Yang Y, Zhang X, Xu X, Wang Y, Han Z (2017) The ethylene response factor AtERF4 negatively regulates the iron deficiency response in Arabidopsis thaliana. PLoS One 12:e0186580
51. Wang Y, Wang Y, Zhu X, Ren Y, Dong H, Duan E, Teng X, Zhao H, Chen R, Chen X, Lei J, Yang H, Tian Y, Chen L, Liu X, Liu S, Jiang L, Wang H, Wan J (2022) Tetrapyrrole biosynthesis pathway regulates plastid-to-nucleus signaling through controlling plastid gene expression in plants. Plant Commun 4: 100411
52. Foyer CH, Hanke G (2022) ROS production and signalling in chloroplasts: cornerstones and evolving concepts. Plant J 111:642
53. Chan KX, Phua SY, Crisp P, McQuinn R, Pogson BJ (2016) Learning the languages of the chloroplast: retrograde signaling and beyond. Annu Rev Plant Biol 67:25–53
54. Shikanai T, Muller-Moule P, Munekage Y, Niyogi KK, Pilon M (2003) PAA1, a P-type ATPase of Arabidopsis, functions in copper transport in chloroplasts. Plant Cell 15: 1333–1346
55. Hantzis LJ, Kroh GE, Jahn CE, Cantrell M, Peers G, Pilon M, Ravet K (2018) A program for iron economy during deficiency targets specific Fe proteins. Plant Physiol 176:596–610
56. Rodriguez-Celma J, Pan IC, Li W, Lan P, Buckhout TJ, Schmidt W (2013) The transcriptional response of Arabidopsis leaves to Fe deficiency. Front Plant Sci 4:276
57. Bashir K, Rasheed S, Kobayashi T, Seki M, Nishizawa NK (2016) Regulating subcellular metal homeostasis: the key to crop improvement. Front Plant Sci 7:1192
58. Bashir K, Ishimaru Y, Shimo H, Nagasaka S, Fujimoto M, Takanashi H, Tsutsumi N, An G, Nakanishi H, Nishizawa NK (2011) The rice mitochondrial iron transporter is essential for plant growth. Nat Commun 2: 322
59. Vigani G, Zocchi G, Bashir K, Philippar K, Briat JF (2013) Signals from chloroplasts and mitochondria for iron homeostasis regulation. Trends Plant Sci 18:305–311
60. Turowski VR, Aknin C, Maliandi MV, Buchensky C, Leaden L, Peralta DA, Busi MV, Araya A, Gomez-Casati DF (2015) Frataxin is localized to both the chloroplast and mitochondrion and is involved in chloroplast

Fe-S protein function in Arabidopsis. PLoS One 10:e0141443

61. Ashykhmina N, Lorenz M, Frerigmann H, Koprivova A, Hofsetz E, Stuhrwohldt N, Flugge UI, Haferkamp I, Kopriva S, Gigolashvili T (2019) PAPST2 plays critical roles in removing the stress signaling molecule 3′--Phosphoadenosine 5′-phosphate from the cytosol and its subsequent degradation in plastids and mitochondria. Plant Cell 31: 231–249
62. Estavillo GM, Crisp PA, Pornsiriwong W, Wirtz M, Collinge D, Carrie C, Giraud E, Whelan J, David P, Javot H, Brearley C, Hell R, Marin E, Pogson BJ (2011) Evidence for a SAL1-PAP chloroplast retrograde pathway that functions in drought and high light signaling in Arabidopsis. Plant Cell 23:3992–4012
63. Quintero FJ, Garciadeblas B, Rodriguez-Navarro A (1996) The SAL1 gene of Arabidopsis, encoding an enzyme with 3′(2′),5′--bisphosphate nucleotidase and inositol polyphosphate 1-phosphatase activities, increases salt tolerance in yeast. Plant Cell 8: 529–537
64. Murguia JR, Belles JM, Serrano R (1996) The yeast HAL2 nucleotidase is an in vivo target of salt toxicity. J Biol Chem 271:29029–29033
65. Chan KX, Mabbitt PD, Phua SY, Mueller JW, Nisar N, Gigolashvili T, Stroeher E, Grassl J, Arlt W, Estavillo GM, Jackson CJ, Pogson BJ (2016) Sensing and signaling of oxidative stress in chloroplasts by inactivation of the SAL1 phosphoadenosine phosphatase. Proc Natl Acad Sci U S A 113:E4567–E4576
66. Gy I, Gasciolli V, Lauressergues D, Morel JB, Gombert J, Proux F, Proux C, Vaucheret H, Mallory AC (2007) Arabidopsis FIERY1, XRN2, and XRN3 are endogenous RNA silencing suppressors. Plant Cell 19:3451–3461
67. Merret R, Descombin J, Juan YT, Favory JJ, Carpentier MC, Chaparro C, Charng YY, Deragon JM, Bousquet-Antonelli C (2013) XRN4 and LARP1 are required for a heat-triggered mRNA decay pathway involved in plant acclimation and survival during thermal stress. Cell Rep 5:1279–1293
68. Zakrzewska-Placzek M, Souret FF, Sobczyk GJ, Green PJ, Kufel J (2010) Arabidopsis thaliana XRN2 is required for primary cleavage in the pre-ribosomal RNA. Nucleic Acids Res 38:4487–4502
69. Van Aken O, Whelan J (2012) Comparison of transcriptional changes to chloroplast and mitochondrial perturbations reveals common and specific responses in Arabidopsis. Front Plant Sci 3:281
70. Balparda M, Armas AM, Estavillo GM, Roschzttardtz H, Pagani MA, Gomez-Casati DF (2020) The PAP/SAL1 retrograde signaling pathway is involved in iron homeostasis. Plant Mol Biol 102:323–337
71. Santos CS, Ozgur R, Uzilday B, Turkan I, Roriz M, Rangel A, Carvalho SMP, Vasconcelos MW (2019) Understanding the role of the antioxidant system and the tetrapyrrole cycle in iron deficiency chlorosis. Plants (Basel) 8
72. Connolly EL, Campbell NH, Grotz N, Prichard CL, Guerinot ML (2003) Overexpression of the FRO2 ferric chelate reductase confers tolerance to growth on low iron and uncovers posttranscriptional control. Plant Physiol 133:1102–1110
73. Balparda M, Armas AM, Gomez-Casati DF, Pagani MA (2021) PAP/SAL1 retrograde signaling pathway modulates iron deficiency response in alkaline soils. Plant Sci 304: 110808
74. Doll S, Kuhlmann M, Rutten T, Mette MF, Scharfenberg S, Petridis A, Berreth DC, Mock HP (2018) Accumulation of the coumarin scopolin under abiotic stress conditions is mediated by the Arabidopsis thaliana THO/-TREX complex. Plant J 93:431–444
75. Mittler R, Vanderauwera S, Gollery M, Van Breusegem F (2004) Reactive oxygen gene network of plants. Trends Plant Sci 9:490–498
76. Olmedo G, Guo H, Gregory BD, Nourizadeh SD, Aguilar-Henonin L, Li H, An F, Guzman P, Ecker JR (2006) ETHYLENE-INSENSITIVE5 encodes a 5′→3′ exoribonuclease required for regulation of the EIN3-targeting F-box proteins EBF1/2. Proc Natl Acad Sci U S A 103:13286–13293
77. Nguyen AH, Matsui A, Tanaka M, Mizunashi K, Nakaminami K, Hayashi M, Iida K, Toyoda T, Nguyen DV, Seki M (2015) Loss of Arabidopsis 5′-3′ exoribonuclease AtXRN4 function enhances heat stress tolerance of plants subjected to severe heat stress. Plant Cell Physiol 56:1762–1772
78. Pornsiriwong W, Estavillo GM, Chan KX, Tee EE, Ganguly D, Crisp PA, Phua SY, Zhao C, Qiu J, Park J, Yong MT, Nisar N, Yadav AK, Schwessinger B, Rathjen J, Cazzonelli CI, Wilson PB, Gilliham M, Chen ZH, Pogson BJ (2017) A chloroplast retrograde signal, 3′--phosphoadenosine 5′-phosphate, acts as a secondary messenger in abscisic acid signaling in stomatal closure and germination. elife 6

79. Rodriguez VM, Chetelat A, Majcherczyk P, Farmer EE (2010) Chloroplastic phosphoadenosine phosphosulfate metabolism regulates basal levels of the prohormone jasmonic acid in Arabidopsis leaves. Plant Physiol 152: 1335–1345
80. Phua SY, Yan D, Chan KX, Estavillo GM, Nambara E, Pogson BJ (2018) The Arabidopsis SAL1-PAP pathway: A case study for integrating chloroplast retrograde, light and hormonal signaling in modulating plant growth and development? Front Plant Sci 9: 1171
81. Romera FJ, Garcia MJ, Alcantara E, Perez-Vicente R (2011) Latest findings about the interplay of auxin, ethylene and nitric oxide in the regulation of Fe deficiency responses by strategy I plants. Plant Signal Behav 6:167–170
82. Pogson BJ, Woo NS, Forster B, Small ID (2008) Plastid signalling to the nucleus and beyond. Trends Plant Sci 13:602–609
83. Susek RE, Ausubel FM, Chory J (1993) Signal transduction mutants of Arabidopsis uncouple nuclear CAB and RBCS gene expression from chloroplast development. Cell 74:787–799
84. Terry MJ, Smith AG (2013) A model for tetrapyrrole synthesis as the primary mechanism for plastid-to-nucleus signaling during chloroplast biogenesis. Front Plant Sci 4:14
85. Woodson JD, Perez-Ruiz JM, Chory J (2011) Heme synthesis by plastid ferrochelatase I regulates nuclear gene expression in plants. Curr Biol 21:897–903
86. Moulin M, McCormac AC, Terry MJ, Smith AG (2008) Tetrapyrrole profiling in Arabidopsis seedlings reveals that retrograde plastid nuclear signaling is not due to Mg-protoporphyrin IX accumulation. Proc Natl Acad Sci U S A 105:15178–15183
87. Tanaka R, Kobayashi K, Masuda T (2011) Tetrapyrrole metabolism in Arabidopsis thaliana. Arabidopsis Book 9:e0145
88. Tanaka R, Tanaka A (2007) Tetrapyrrole biosynthesis in higher plants. Annu Rev Plant Biol 58:321–346
89. Moulin M, Smith AG (2005) Regulation of tetrapyrrole biosynthesis in higher plants. Biochem Soc Trans 33:737–742
90. Watanabe N, Che FS, Iwano M, Takayama S, Yoshida S, Isogai A (2001) Dual targeting of spinach protoporphyrinogen oxidase II to mitochondria and chloroplasts by alternative use of two in-frame initiation codons. J Biol Chem 276:20474–20481
91. Chow KS, Singh DP, Roper JM, Smith AG (1997) A single precursor protein for ferrochelatase-I from Arabidopsis is imported in vitro into both chloroplasts and mitochondria. J Biol Chem 272:27565–27571
92. Larkin RM, Alonso JM, Ecker JR, Chory J (2003) GUN4, a regulator of chlorophyll synthesis and intracellular signaling. Science 299: 902–906
93. Mochizuki N, Brusslan JA, Larkin R, Nagatani A, Chory J (2001) Arabidopsis genomes uncoupled 5 (GUN5) mutant reveals the involvement of Mg-chelatase H subunit in plastid-to-nucleus signal transduction. Proc Natl Acad Sci U S A 98:2053–2058
94. Strand A, Asami T, Alonso J, Ecker JR, Chory J (2003) Chloroplast to nucleus communication triggered by accumulation of Mg-protoporphyrinIX. Nature 421:79–83
95. Mochizuki N, Tanaka R, Tanaka A, Masuda T, Nagatani A (2008) The steady-state level of Mg-protoporphyrin IX is not a determinant of plastid-to-nucleus signaling in Arabidopsis. Proc Natl Acad Sci U S A 105:15184–15189
96. Voigt C, Oster U, Bornke F, Jahns P, Dietz KJ, Leister D, Kleine T (2010) In-depth analysis of the distinctive effects of norflurazon implies that tetrapyrrole biosynthesis, organellar gene expression and ABA cooperate in the GUN-type of plastid signalling. Physiol Plant 138:503–519
97. Martin M, Colman MJ, Gomez-Casati DF, Lamattina L, Zabaleta EJ (2009) Nitric oxide accumulation is required to protect against iron-mediated oxidative stress in frataxin-deficient Arabidopsis plants. FEBS Lett 583:542–548
98. Busi MV, Maliandi MV, Valdez H, Clemente M, Zabaleta EJ, Araya A, Gomez-Casati DF (2006) Deficiency of Arabidopsis thaliana frataxin alters activity of mitochondrial Fe-S proteins and induces oxidative stress. Plant J 48:873–882
99. Maliandi MV, Busi MV, Clemente M, Zabaleta EJ, Araya A, Gomez-Casati DF (2007) Expression and one-step purification of recombinant Arabidopsis thaliana frataxin homolog (AtFH). Protein Expr Purif 51: 157–161
100. Buchensky C, Sanchez M, Carrillo M, Palacios O, Capdevila M, Dominguez-Vera JM, Busi MV, Atrian S, Pagani MA, Gomez-Casati DF (2017) Identification of two frataxin isoforms in Zea mays: structural and functional studies. Biochimie 140:34–47
101. Lesuisse E, Santos R, Matzanke BF, Knight SA, Camadro JM, Dancis A (2003) Iron use

for haeme synthesis is under control of the yeast frataxin homologue (Yfh1). Hum Mol Genet 12:879–889
102. Yoon T, Cowan JA (2004) Frataxin-mediated iron delivery to ferrochelatase in the final step of heme biosynthesis. J Biol Chem 279: 25943–25946
103. Napoli E, Morin D, Bernhardt R, Buckpitt A, Cortopassi G (2007) Hemin rescues adrenodoxin, heme a and cytochrome oxidase activity in frataxin-deficient oligodendroglioma cells. Biochim Biophys Acta 1772:773–780
104. Maliandi MV, Busi MV, Turowski VR, Leaden L, Araya A, Gomez-Casati DF (2011) The mitochondrial protein frataxin is essential for heme biosynthesis in plants. FEBS J 278:470–481
105. Masuda T, Suzuki T, Shimada H, Ohta H, Takamiya K (2003) Subcellular localization of two types of ferrochelatase in cucumber. Planta 217:602–609
106. Cornah JE, Roper JM, Pal Singh D, Smith AG (2002) Measurement of ferrochelatase activity using a novel assay suggests that plastids are the major site of haem biosynthesis in both photosynthetic and non-photosynthetic cells of pea (Pisum sativum L.). Biochem J 362:423–432
107. Armas AM, Balparda M, Terenzi A, Busi MV, Pagani MA, Gomez-Casati DF (2019) Ferrochelatase activity of plant frataxin. Biochimie 156:118–122
108. Busi MV, Gomez-Casati DF (2012) Exploring frataxin function. IUBMB Life 64:56–63
109. Monfort B, Want K, Gervason S, D'Autreaux B (2022) Recent advances in the elucidation of Frataxin biochemical function open novel perspectives for the treatment of Friedreich's ataxia. Front Neurosci 16:838335
110. Salome PA, Oliva M, Weigel D, Kramer U (2013) Circadian clock adjustment to plant iron status depends on chloroplast and phytochrome function. EMBO J 32:511–523
111. Chen YY, Wang Y, Shin LJ, Wu JF, Shanmugam V, Tsednee M, Lo JC, Chen CC, Wu SH, Yeh KC (2013) Iron is involved in the maintenance of circadian period length in Arabidopsis. Plant Physiol 161:1409–1420
112. Tissot N, Przybyla-Toscano J, Reyt G, Castel B, Duc C, Boucherez J, Gaymard F, Briat JF, Dubos C (2014) Iron around the clock. Plant Sci 224:112–119
113. Beinert H (2000) Iron-sulfur proteins: ancient structures, still full of surprises. J Biol Inorg Chem 5:2–15
114. Braymer JJ, Lill R (2017) Iron-sulfur cluster biogenesis and trafficking in mitochondria. J Biol Chem 292:12754–12763
115. Ranatunga W, Gakh O, Galeano BK, Smith DY, Soderberg CA, Al-Karadaghi S, Thompson JR, Isaya G (2016) Architecture of the yeast mitochondrial iron-sulfur cluster assembly machinery: the sub-complex formed by the iron donor, Yfh1 protein, and the scaffold, Isu1 protein. J Biol Chem 291:10378–10398
116. Balk J, Pilon M (2011) Ancient and essential: the assembly of iron-sulfur clusters in plants. Trends Plant Sci 16:218–226
117. Li L, Miao R, Bertram S, Jia X, Ward DM, Kaplan J (2012) A role for iron-sulfur clusters in the regulation of transcription factor Yap5-dependent high iron transcriptional responses in yeast. J Biol Chem 287:35709–35721
118. Poor CB, Wegner SV, Li H, Dlouhy AC, Schuermann JP, Sanishvili R, Hinshaw JR, Riggs-Gelasco PJ, Outten CE, He C (2014) Molecular mechanism and structure of the Saccharomyces cerevisiae iron regulator Aft2. Proc Natl Acad Sci U S A 111:4043–4048
119. Anderson CP, Shen M, Eisenstein RS, Leibold EA (2012) Mammalian iron metabolism and its control by iron regulatory proteins. Biochim Biophys Acta 1823:1468–1483
120. Arnaud N, Ravet K, Borlotti A, Touraine B, Boucherez J, Fizames C, Briat JF, Cellier F, Gaymard F (2007) The iron-responsive element (IRE)/iron-regulatory protein 1 (IRP1)-cytosolic aconitase iron-regulatory switch does not operate in plants. Biochem J 405:523–531
121. Anderson PR, Kirby K, Hilliker AJ, Phillips JP (2005) RNAi-mediated suppression of the mitochondrial iron chaperone, frataxin, in Drosophila. Hum Mol Genet 14:3397–3405
122. Leaden L, Pagani MA, Balparda M, Busi MV, Gomez-Casati DF (2016) Altered levels of AtHSCB disrupts iron translocation from roots to shoots. Plant Mol Biol 92:613–628
123. Bitto E, Bingman CA, Bittova L, Kondrashov DA, Bannen RM, Fox BG, Markley JL, Phillips GN Jr (2008) Structure of human J-type co-chaperone HscB reveals a tetracysteine metal-binding domain. J Biol Chem 283: 30184–30192
124. Armas AM, Balparda M, Turowski VR, Busi MV, Pagani MA, Gomez-Casati DF (2019) Altered levels of mitochondrial NFS1 affect cellular Fe and S contents in plants. Plant Cell Rep 38:981–990
125. Turowski VR, Busi MV, Gomez-Casati DF (2012) Structural and functional studies of

the mitochondrial cysteine desulfurase from Arabidopsis thaliana. Mol Plant 5:1001–1010

126. Forieri I, Sticht C, Reichelt M, Gretz N, Hawkesford MJ, Malagoli M, Wirtz M, Hell R (2017) System analysis of metabolism and the transcriptome in Arabidopsis thaliana roots reveals differential co-regulation upon iron, sulfur and potassium deficiency. Plant Cell Environ 40:95–107
127. Forieri I, Wirtz M, Hell R (2013) Toward new perspectives on the interaction of iron and sulfur metabolism in plants. Front Plant Sci 4:357
128. Zuchi S, Watanabe M, Hubberten HM, Bromke M, Osorio S, Fernie AR, Celletti S, Paolacci AR, Catarcione G, Ciaffi M, Hoefgen R, Astolfi S (2015) The interplay between sulfur and iron nutrition in tomato. Plant Physiol 169:2624–2639
129. Vigani G, Briat JF (2015) Impairment of respiratory chain under nutrient deficiency in plants: does it play a role in the regulation of iron and sulfur responsive genes? Front Plant Sci 6:1185
130. Kobayashi T, Nishizawa NK (2015) Intracellular iron sensing by the direct binding of iron to regulators. Front Plant Sci 6:155
131. Hirabayashi K, Yuda E, Tanaka N, Katayama S, Iwasaki K, Matsumoto T, Kurisu G, Outten FW, Fukuyama K, Takahashi Y, Wada K (2015) Functional dynamics revealed by the structure of the SufBCD complex, a novel ATP-binding cassette (ABC) protein that serves as a scaffold for iron-sulfur cluster biogenesis. J Biol Chem 290:29717–29731
132. Hu X, Kato Y, Sumida A, Tanaka A, Tanaka R (2017) The SUFBC2 D complex is required for the biogenesis of all major classes of plastid Fe-S proteins. Plant J 90:235–248
133. Tanaka R, Tanaka A (2011) Chlorophyll cycle regulates the construction and destruction of the light-harvesting complexes. Biochim Biophys Acta 1807:968–976
134. Zandalinas SI, Song L, Sengupta S, McInturf SA, Grant DG, Marjault HB, Castro-Guerrero NA, Burks D, Azad RK, Mendoza-Cozatl DG, Nechushtai R, Mittler R (2020) Expression of a dominant-negative AtNEET-H89C protein disrupts iron-sulfur metabolism and iron homeostasis in Arabidopsis. Plant J 101:1152–1169
135. Tamir S, Paddock ML, Darash-Yahana-Baram M, Holt SH, Sohn YS, Agranat L, Michaeli D, Stofleth JT, Lipper CH, Morcos F, Cabantchik IZ, Onuchic JN, Jennings PA, Mittler R, Nechushtai R (2015) Structure-function analysis of NEET proteins uncovers their role as key regulators of iron and ROS homeostasis in health and disease. Biochim Biophys Acta 1853:1294–1315
136. Nechushtai R, Conlan AR, Harir Y, Song L, Yogev O, Eisenberg-Domovich Y, Livnah O, Michaeli D, Rosen R, Ma V, Luo Y, Zuris JA, Paddock ML, Cabantchik ZI, Jennings PA, Mittler R (2012) Characterization of Arabidopsis NEET reveals an ancient role for NEET proteins in iron metabolism. Plant Cell 24:2139–2154
137. Mittler R, Darash-Yahana M, Sohn YS, Bai F, Song L, Cabantchik IZ, Jennings PA, Onuchic JN, Nechushtai R (2019) NEET proteins: A new link between iron metabolism, reactive oxygen species, and cancer. Antioxid Redox Signal 30:1083–1095

Chapter 12

Functional Analysis of Chloroplast Iron Uptake and Homeostasis

Helga Zelenyánszki and Ádám Solti

Abstract

Iron has a crucial role in plastid biology. Iron is a required cofactor for the operation of the photosynthetic functions and other metabolic pathways. Despite the importance of the iron homeostasis in chloroplasts, the functional analysis of the plastidial iron uptake and homeostasis still lack a consensus methodology. Here, we describe a sequence of subsequent techniques that can be applied in functional characterization of proteins involved in iron uptake and incorporation into chloroplasts as well as of the non-transport protein members of the chloroplast iron homeostasis. Since the ferrous iron ligation of bathophenantroline disulfonate is specific and not disrupted by the presence of other transition metals, it offers a simple way for iron quantification both in solubilized chloroplast samples as well as in ferric chelate reductase activity measurements.

Key words Bathophenantroline disulfonate, Chloroplast envelope membrane, Ferric chelate reductase, Gradient centrifugation, Mössbauer spectroscopy, Spectrophotometry

1 Introduction

Although iron (Fe) plays a crucial role in chloroplast biology and photosynthesis, the information on the mechanisms of chloroplast Fe acquisition, incorporation, and liberation is scarce, thus their functional analysis is required. In photosynthetically active chloroplasts, Fe primarily incorporates into the photosynthetic apparatus [1, 2] in the form of heme groups and Fe–S clusters. Nevertheless, these cofactors, among other cofactors of the oxygen transfer reactions operated by cytochrome P450 monooxygenases and 2-oxoglutarate-dependent dioxygenases, are also involved in the nitrogen fixation and DNA synthesis via the activity of the ribonucleotide reductase [3]. Moreover, Fe also serves as a cofactor for a wide range of antioxidative defense enzymes [4]. Shortage of Fe in the chloroplasts leads to impaired biogenesis and thus dysfunction of the photosynthetic apparatus. Therefore, perturbed plastidial Fe

Jeeyon Jeong (ed.), *Plant Iron Homeostasis: Methods and Protocols*, Methods in Molecular Biology, vol. 2665,
https://doi.org/10.1007/978-1-0716-3183-6_12,

homeostasis is associated with reduced growth and biomass accumulation [5]. In the past 15 years, multiple chloroplast Fe homeostasis elements have been described [6]. However, functional characterization of the protein members of this system is required, especially in the native plastidial environment. Multiple functional characterization techniques require an isolated chloroplast system that lacks contamination of other cell constituents, such as the mitochondria. Since purified chloroplasts are highly sensitive objects, special care is needed to avoid any damages during the functional activity studies. The photosynthetic machinery is also capable to reduce various ferric Fe salts [7], thus a special care is needed to avoid artifacts generated by the light reactions of the photosynthetic apparatus of damaged chloroplasts/thylakoid membrane fragments. Characterization studies may also require isolated membrane systems. Since chloroplasts are bordered by a double envelope membrane system, a proper characterization starts with the separation and isolation of the envelope membranes of purified chloroplasts. Here we demonstrate a setup of protocols that enable a holistic characterization of an arbitrary chloroplast localized Fe homeostasis element that affect either the transport or the proper incorporation of Fe in chloroplasts (Fig. 1).

2 Materials

Prepare all solutions using deionized water (conductivity less than 2 μS cm^{-1}). Take special care to avoid residual Fe contamination in the deionized water that can arise from metal pieces of any system. In any case of unknown water quality, we suggest a preliminary elemental analysis of the water. Also take care to avoid Fe contamination in the chemicals applied in the studies. Certain widely used gradient materials (e.g., Percoll, Ficoll) are often sources of such Fe contaminations. Since Fe complexes are photoactive, keep Fe complex solution covered / in darkness, and use them in a short time or keep frozen at −20 °C. Avoid detergent residues from cleaning the glassware that is used to prepare and store the solutions, because chloroplast and envelopes are extremely sensitive to solubilization by detergents.

2.1 Plants and Materials for Plant Growth

1. Sugar beet (*Beta vulgaris* L.) or oilseed rape (*Brassica napus* L.).
2. Vermiculite for germination. Alternatively, seeds can be also germinated on wet filter paper in Petri dish.
3. Controlled environment or growth chamber (*see* **Note 1**).
4. 12 l plastic bucket
5. Styrofoam sheets of 1.5 cm thickness.

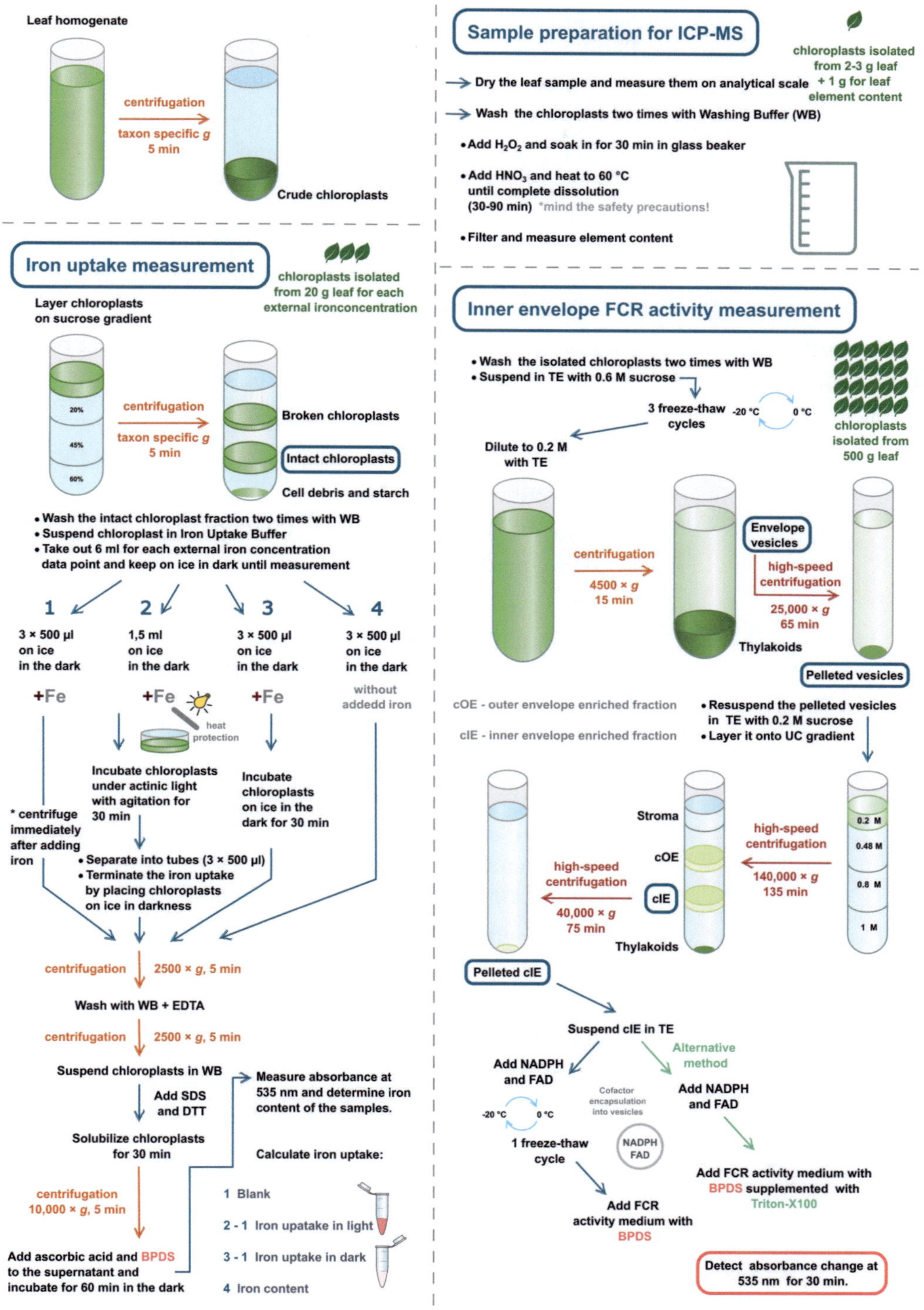

Fig. 1 Schematic overview of the methods used for characterization of iron status of plants. Crude chloroplast pellet obtained by centrifugation of the filtrated leaf homogenate can be used is the analysis of elements in the chloroplasts (*see* Subheading 3.4). To perform a chloroplast iron uptake study (*see* Subheading 3.6) the purification of intact class I chloroplasts by gradient centrifugation is essential. The iron uptake of chloroplasts is a light driven process. By comparing the iron content of a plastid suspension prior and after its illumination in the presence of iron compounds, the amount of iron that was taken up into the chloroplasts can be

6. Sponge cuts of 1 cm width for planting and cultivating plant material.
7. Black garden foil to cover the styrofoam sheets to prevent algal colonization.
8. Modified quarter-strength Hoagland's solution: 1.25 mM $Ca(NO_3)_2$, 1.25 mM KNO_3, 0.5 mM $MgSO_4$, 0.25 mM KH_2PO_4, 11.56 μM H_3BO_3, 4.6 μM $MnCl_2$, 0.19 μM $ZnSO_4$, 0.12 μM Na_2MoO_4, 0.08 μM $CuSO_4$ supplied with the proper form and concentration of Fe (*see* **Notes 2** and **3**).

2.2 Chloroplast Isolation and Purification

1. Fully developed, non-senescent leaves of sugar beet or oilseed rape (*see* **Note 4**).
2. Ice.
3. Pre-chilled (to 4 °C) laboratory blender (e.g., Waring).
4. Pharmaceutical gauze sheets.
5. Miracloth™ sheets, pore size of 20–25 μm, cut to approx. 15 × 15 cm size.
6. Funnel.
7. Chloroplast Isolation Buffer (IB): 50 mM HEPES-KOH (pH 7.0), 330 mM sorbitol, 2 mM EDTA (added from 200 mM stock of K-EDTA), 2 mM $MgCl_2$ (added from 200 mM stock of $MgCl_2 \times 6H_2O$), 0.1% (w/V) bovine serum albumin (BSA), 0.1% (w/V) sodium ascorbate. Adjust pH with 10% (w/V) KOH (*see* **Note 5**).
8. Refrigerated centrifuge with a swing-out rotor (for chloroplast isolation) and a fixed angle rotor (for chlorophyll content determination), and precooled centrifuge tubes.
9. Washing buffer (WB): 50 mM HEPES-KOH (pH 7.0), 330 mM sorbitol, 2 mM $MgCl_2$.
10. Fine paintbrush of size 2–3 made of natural hair or wide opening Pasteur pipette.
11. Gradient buffer (GB): 50 mM HEPES-KOH (pH 7.0), 330 mM sorbitol, 2 mM $MgCl_2$, supplemented with 20, 45, or 60% (w/V) sucrose (*see* **Note 6**).

Fig. 1 (continued) determined by a simple measurement using spectrophotometry, after washing, solubilizing the chloroplasts and enable the complex formation of bathophenanthrolinedisulfonic acid disodium salt with ferrous iron under a reducing environment. The isolation of chloroplast envelope membranes (*see* Subheading 3.2) also starts with the isolation of chloroplasts. The separation of chloroplast envelope membranes requires high-speed gradient centrifugation. Chloroplast inner envelope membrane vesicles obtained by this protocol can be subjected to ferric chelate reductase assays (*see* Subheading 3.7)

12. 60/45/20% GB gradient in a 50 ml conical tube for gradient centrifugation: Prepare the gradient by carefully layering 5/6/5 ml of cold GB with 60/45/20% (w/V) sucrose into a 45 ml tube and keep on ice until overlay the crude chloroplast fraction.
13. Tricine-buffered acetone: 80% (V/V) acetone, 5 mM Tricin-KOH (pH 7.8).
14. Spectrophotometer for total chlorophyll determination.

2.3 Determining Chloroplast Suspension Purity and Intactness

1. Polyacrylamide gel electrophoresis and horizontal transfer (western blotting) system.
2. Gel detection system and densitometry evaluation software (*see* **Note 7**).
3. Precast SDS gels or (12% continuous or 10–18% gradient) SDS gels prepared as in [8].
4. Protein solubilization buffer (PSB): 62.5 mM Tris-HCl (pH 6.8), 2% (w/V) SDS, 2% (w/V) dithiothreitol (DTT), 8.7% (V/V) glycerol and 0.001% (w/V) bromophenol blue.
5. Nitrocellulose membrane, blot filter paper.
6. Transfer buffer: 192 mM glycine, 25 mM Tris-HCL (pH 8.3) 20% (V/V) methanol, 0.02% (w/V) SDS.
7. Tris-buffered saline (TBS): 20 mM Tris, 500 mM NaCl, (pH 7.5) and TBS supplemented with 0.05% (V/V) Tween-20 (TBST).
8. Gelatin or non-fat milk powder for blocking.
9. Antibodies against non-plastid cell constituents (*see* **Note 8**).
10. Antibodies against chloroplast constituents:
    (a) Chloroplast inner envelope (cIE) marker: chloroplast triose-phosphate translocator (TPT) or inner envelope protein 37 (IEP37).
    (b) Chloroplast outer envelope (cOE) marker: a protein component of the cOE translocon complex: TOC75.
    (c) Thylakoid marker: light-harvesting complex apoproteins (apoLHC).
    (d) Chloroplast stroma marker: RbcL.
11. Secondary antibodies.
12. Appropriate development solutions according to the applied secondary antibodies.

2.4 Iron Content Determination

1. Suspension containing intact chloroplasts.
2. Fe uptake medium: WB supplemented with the proper form and concentration of Fe (*see* **Note 9**). Medium contains 50 mM

HEPES-KOH (pH 7.0), 330 mM sorbitol, 2 mM $MgCl_2$, Fe source in complexed form stable at pH 7.0.

3. Actinic light source of 120 μmol photon m^{-2} s^{-1} intensity (*see* **Note 10**).
4. Surface-bound Fe removal washing solution: WB supplemented with 2 mM EDTA (*see* **Note 11**).
5. Chloroplast solubilization: 10% (w/V) SDS, 10% (w/V) DTT.
6. Fe content measurement solution: 1 mM ascorbic acid, 3 mM bathophenanthrolinedisulfonic acid disodium salt (BPDS) (*see* **Note 12**).
7. Precooled (4 °C) centrifuge, swing-out rotor mounted, and centrifuge, angle rotor mounted, operating at room temperature.
8. Spectrophotometer to measure absorbance at 800 and 535 nm.

2.5 Sample Preparation for Element Analysis by ICP-MS

1. Analytical scale for dry weight determination.
2. Oven set at 60 °C.
3. 50 ml beakers, watch glass.
4. Analytical grade 30% (V/V) H_2O_2.
5. Analytical grade cc. HNO_3.
6. Deionized water (conductivity less than 2 μS cm^{-1}).
7. Laboratory fume hood, cc. HNO_3-resistant gloves, safety goggles.
8. Hot plate capable of heating up to 60 °C.
9. MN 640 W paper to filter the digested solutions.

2.6 Mössbauer Spectroscopy

1. Radiation source: ^{57}Co that has a high intensity (within one year of usage).
2. Cryostat operating with liquid nitrogen.
3. Mössbauer spectrometer and evaluating software.
4. Plant material cultivated on nutrient solution supplemented with, or chloroplast suspension that was treated by ^{57}Fe source.

2.7 Chloroplast Envelope Isolation

1. Tricine-EDTA (TE) buffer: 10 mM Tricine-KOH (pH 7.8), 2 mM EDTA.
2. Storage Buffer (SB): TE buffer supplemented with 0.6 M sucrose.
3. High-speed centrifugation Gradient Buffers (HGB): TE buffer supplemented with 1 M, 0.8 M or 0.46 M sucrose.
4. −20 °C freezer.

5. Refrigerated centrifuge, swing-out rotor mounted; precooled centrifuge tubes.
6. Ultracentrifuge, swing-out rotor capable of holding 14 ml tubes. The rotor should be able to speed up to 35,000 rpm (approx. 140,000 × *g* medium force in Beckman Sw40Ti operating in L7 ultracentrifuge).

2.8 Ferric Chelate Reductase Activity

1. Chloroplast inner envelope vesicle suspension in TE buffer.
2. FCR activity medium: 50 mM HEPES-KOH (pH 7.0), 330 mM sorbitol, 2 mM $MgCl_2$.
3. 3 mM BPDS stock solution
4. 10 mM NADPH, 10 mM FAD, and 10 mM Fe (III)-EDTA stock solutions
5. −20 °C freezer
6. Spectrophotometer to measure absorbance at 800 and 535 nm.

2.8.1 Alternative Method to Measure Ferric Chelate Reductase Activity

1. Chloroplast inner envelope vesicle suspension in TE supplemented.
2. 1% (V/V) Triton X-100 for membrane solubilization (*see* **Note 13**)
3. 10 mM NADPH, 10 mM FAD, and 10 mM Fe (III)-EDTA stock solutions
4. Spectrophotometer to measure absorbance at 800 and 535 nm.

2.9 Transient Expression System to Prove Chloroplast Localization of Iron Homeostasis Elements

1. Common bean (*Phaseolus vulgaris* L., *see* **Note 14**) seedlings of 10–20 days old.
2. *Rhizobium radiobacter* (formerly: *Agrobacterium tumefaciens*) AGL1 strain.
3. Plant transformation vector with a constitutive promoter and GFP tag (e.g., pCAMBIA1302).
4. Handheld UV or blue lamp, orange optical filter, and/or orange safety goggles.
5. Luria-Bertani (LB) broth: 5 mg ml^{-1} tryptone, 5 mg ml^{-1} yeast extract, 10 mg ml^{-1} NaCl.
6. 25 mg ml^{-1} rifampicin stock solution; 1000 times concentrated stock of the respective antibiotic specific for the used plant transformation vector
7. Agro-infiltration medium: 10 mM $MgCl_2$, 10 mM MES (pH 5.6).
8. Spectrophotometer to measure optical density (OD) at 600 nm.

2.10 Protoplast Isolation

1. Protoplast Buffer (PB): 500 mM sorbitol, 2 mM MES-KOH (pH 5.6), 20 mM KCl.
2. Protoplast Isolation Buffer: PB supplemented with 2% (w/V) cellulase (Onozuka R-10), 1% (w/V) macerozyme (Onozuka R-10), 10 mM $CaCl_2$, 2 mM DTT, 0.2% (w/V) BSA (*see* **Note 15**).
3. Protoplast Washing Buffer: PB supplemented with 0.5% (w/V) BSA (*see* **Note 16**).
4. Blades, small beaker (⌀ = 2–3 cm), parafilm, rotary shaker.
5. Gauze sheets, cell strainer (100 μm).
6. Preparative centrifuge, swing-out rotor mounted.

2.11 Protoplast Immunofluorescence Staining

1. Preparative centrifuge, swing-out rotor mounted.
2. 2-times concentrated protoplast fixation medium: 8% (w/V) paraformaldehyde, 2 mM MES, 20 mM KCL, 0.5 M sorbitol (*see* **Note 17**).
3. Immunofluorescence Washing Buffer (IFWB): PBS supplemented with 0.5% (w/V) BSA.
4. Immunofluorescence Blocking Buffer (IBB): PBS supplemented with 5% (w/V) BSA and 0.5% (V/V) Tween-20.
5. Anti-GFP primary antibody and fluorescent secondary antibody. (*see* **Note 18**).
6. Microscope slides, coverslips, anti-fade fluorescent mounting medium, optional: DAPI to stain the nucleus and BioxMLYellow (Bioxol Ltd.) to dye the membranes (*see* **Note 19**).
7. Confocal laser scanning microscope.

3 Methods

3.1 Chloroplast Isolation and Purification

1. Grow sugar beet or oilseed rape plants in hydroponics according to the appropriate Fe nutrition scheme of the study.
2. Two days prior to the chloroplast isolation the model plants should be placed in the darkness to decrease the starch content of the chloroplasts.
3. All procedures are carried out in cold conditions; thus, it is recommended to keep the buffers and centrifuge tubes on ice.
4. Collect the leaves on ice and cut them into 1 cm stripes directly into the chloroplast isolation buffer. The ideal leaf-to-buffer ratio is 10:40 (g:ml).

5. Homogenize the cut leaves for 3 s in the blender (*see* **Note 20**), then filter the homogenate through 4 layers of gauze and one layer of Miracloth using a funnel to remove cell debris.
6. Take 500 μl sample of the homogenate for the chloroplast intactness measurements, snap freeze in liquid nitrogen, and store the samples at −80 °C for later use.
7. Pellet the chloroplasts from the crude homogenate by centrifugation. (*see* **Notes 21** and **22**). A few examples of effective pelleting conditions are:
   (a) Sugar beet, optimum nutrition: 1,500 × *g*, 5 min.
   (b) Sugar beet, Fe deficient: 2,000 × *g*, 5 min.
   (c) Oilseed rape, optimum nutrition 1,600 × *g*, 5 min.
   (d) Oilseed rape, Fe deficient: 2,500 × *g*, 5 min.
8. Remove the supernatant by pouring it down (*see* **Note 23**). Suspend the chloroplast pellet gently in 0.5–1 ml WB using a fine natural brush or Pasteur pipette.
9. Dilute the chloroplast suspension with WB until it is a light green color and pellet it again applying the taxon-specific centrifugation scheme.
   (a) To proceed towards chloroplast envelope isolation, repeat the washing step (Step 6.), then resuspend the pellet in TE buffer supplemented with 0.6 M sucrose.
   (b) To proceed towards purified chloroplast suspension, resuspend the chloroplast in 1 ml WB and lay it on a precooled 60/45/20% (w/V) sucrose gradient (GB) gently (Fig. 1).
10. Gradient centrifugation is performed in a swing-out rotor at 2,000 × *g* for 20 min (*see* **Note 24).**
11. Collect the intact chloroplast fraction with a Pasteur pipette at first. Avoid the cross-contamination of fractions with class II chloroplasts originating from the top of the gradient.
12. Dilute the suspension five to tenfold with WB then pellet the chloroplasts at 2,500 × *g* for 5 min (*see* **Note 25).**
13. Suspend the chloroplast pellet in TE supplemented with 0.6 M sucrose (chloroplast envelope isolation) or in WB (further work with intact chloroplasts).
14. Determine the chloroplast density of the suspension in aliquots 100-times diluted by WB using a Bürker chamber in a light or epifluorescence microscope mounted with adequate filters for chlorophyll fluorescence detection ($Ex_{480/30}$, Em_{600lp} or with Cy5 filter set) and 40× objective mounted (*see* **Note 26**). Alternatively, chloroplasts can also be counted using phase-contrast microscopy (Fig. 3).

15. To determine total chlorophyll concentration, take 20 μl aliquots of the chloroplast suspension and dilute the aliquot 100 times by 80% (V/V) Tricine-buffered acetone.
16. Remove the precipitated proteins and the starch pellet by centrifugation in an angle rotor at 4 °C, 10,000 × *g*, for 10 min.
17. Determine the total chlorophyll content in a UV-VIS spectrophotometer using the absorption coefficients of e.g., [9].

3.2 Isolation of Chloroplast Envelope Membranes

Chloroplast envelope membranes from sugar beet and oilseed rape models can be isolated by the modified protocol based on Keegstra and Yousif and Froelich et al. [10, 11] (Fig. 1).

1. Resuspend chloroplasts in TE supplemented with 0.6 M sucrose in three freeze-thaw cycles (−20 °C/0 °C).
2. Repeat three freeze-thaw cycles (−20 °C/0 °C): Freeze chloroplast suspension by placing 1.5 ml aliquots in microcentrifuge tubes for 20–30 min, and then place the suspension into ice containing water bath immediately after freezing for approximately 30 min.
3. Following the third freeze-thaw cycle, dilute the suspension threefold with TE buffer to achieve a final sucrose concentration of 0.2 M and incubate the suspension for an hour on ice (*see* **Note 27**).
4. Remove most of the thylakoid membranes by centrifugation at 4,500 × *g* for 15 min in swing-out rotor.
5. Collect the supernatant by carefully avoiding thylakoid contamination (*see* **Note 28**). If necessary, the supernatant enriched in chloroplast envelope membranes can be stored at −20 °C overnight. Pellet the membranes at 25,000 × *g* for 65 min using a swing-out rotor (Sw40Ti or equivalent) in high-speed centrifuge.
6. Remove the supernatant and suspend the pellet in TE buffer supplemented with 0.2 M sucrose. The pellet may be green due to presence of thylakoid membranes. In case of using 11 ml ultracentrifuge tube, for resuspension, use maximum 0.5 ml of TE to each tube.
7. Prepare a sucrose gradient in transparent ultracentrifuge tubes (e.g., Beckman Ultra-Clear Tube) for high-speed gradient centrifugation on ice: 3 ml TE supplemented with 1.0 M sucrose ($\rho = 1.13$ g ml^{-1}), 3 ml TE supplemented with 0.8 M sucrose ($\rho = 1.10$ g ml^{-1}), and 3 ml TE supplemented with 0.46 M sucrose ($\rho = 1.06$ g ml^{-1}) are layered in 14 ml ultracentrifuge tube. The gradient is overlayered with 2 ml chloroplast envelope membranes enriched supernatant (this fraction is in TE containing 0.2 M sucrose ($\rho = 1.03$ g ml^{-1}).

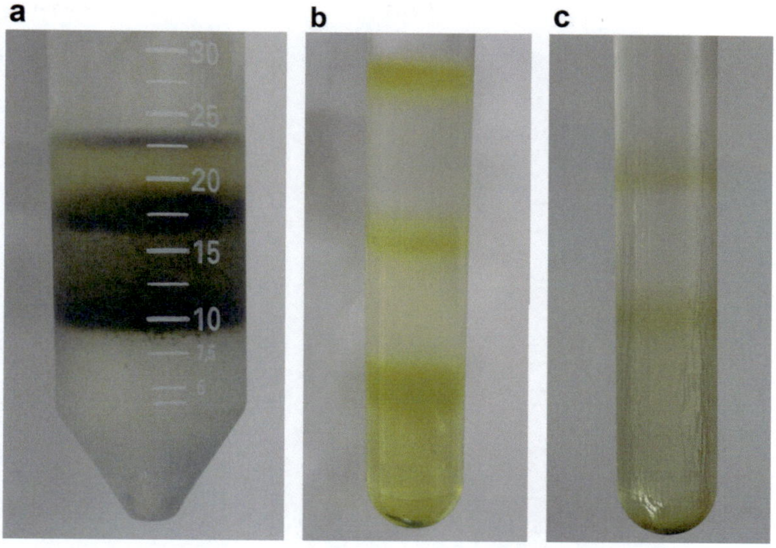

Fig. 2 Result of a gradient centrifugation of a crude chloroplast suspension performed in swing-out rotor at 2,000 × *g* for 20 min (**a**) and results of a high-speed gradient centrifugation of a crude chloroplast envelope membrane suspension obtained from iron deficient plants (**b**) and plants grown under optimum iron nutrition (**c**), centrifugation performed at 140,000 × *g* for 135 min in a swing-out rotor. During the purification of chloroplasts (**a**) intact, class I chloroplasts remain on the interface between the 45 and 60% (w/V) sucrose containing GB, while the broken, class II chloroplasts remain above, at the interface of the 20% and 45% (w/V) sucrose containing GB steps. As the result of a high-speed centrifugation of a mixed chloroplast envelope suspension (**b**, **c**) inner chloroplast envelope membrane vesicles remain at the interface between the 1.0 M and 0.8 M sucrose containing TE. A fraction enriched in chloroplast outer envelope membrane vesicles remains above, at the interface between the 0.8 M and 0.46 M sucrose containing TE. In the top, chloroplast stroma proteins, together with broken thylakoid membranes remains at the interface between the 0.46 M and 0.2 M sucrose containing TE. Thylakoid membranes that remained in the crude chloroplast envelope membrane suspension sediment to the bottom of the 1.0 M gradient step

8. Perform high-speed gradient centrifugation at 140,000 × *g* for 135 min in a swing-out rotor.
9. Collect the separated membrane vesicle fractions, preferentially starting at 1.0/0.8 M sucrose interphase (Fig. 2b, c):
   (a) Chloroplast inner envelope (cIE) vesicle fraction at the 1.0/0.8 M sucrose interphase.
   (b) Chloroplast envelope membrane fraction enriched in outer envelope (cOE) at the 0.8/0.46 M sucrose interface.

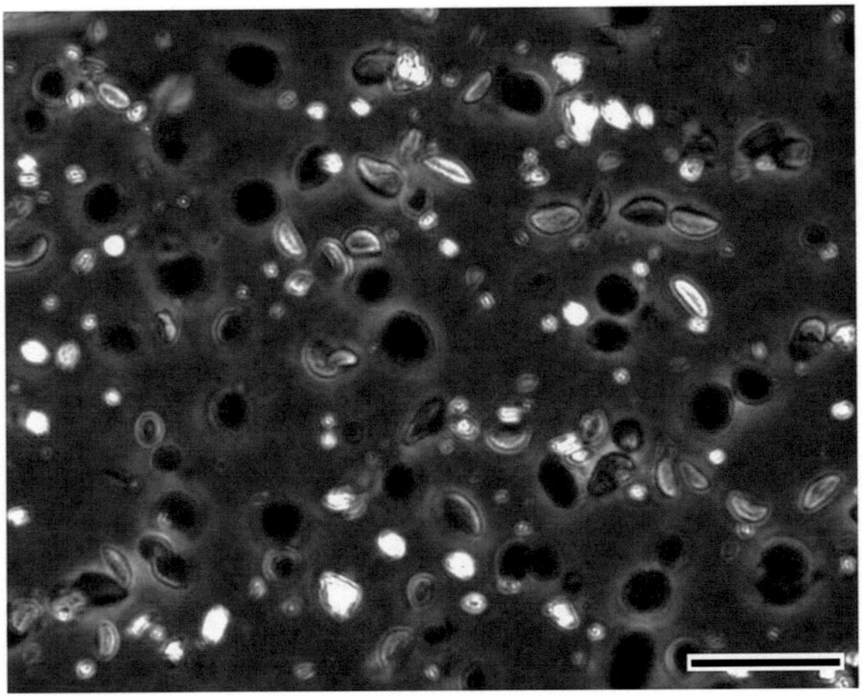

Fig. 3 Purified chloroplast suspension obtained from sugar beet leaves examined by phase-contract microscopy. In the micrograph, damaged class II chloroplasts appear as dark particles. Intact class I chloroplasts appear as grey-light particles, retaining their typical appearance. Small light particles are starch granules. Bar is equal to 20 μm

(c) Broken thylakoids and other chloroplast stroma fractions at the 0.46/0.2 M sucrose interface.

(d) After the removal of the gradient, the thylakoid fraction from the bottom of the ultracentrifuge tube can be collected by resuspending it in TE buffer supplemented with 0.2 M sucrose.

10. Dilute the isolated envelope vesicles with TE buffer supplemented with 0.2 M sucrose and pellet the membranes using high-speed centrifugation at 40,000 × *g* for 75 min in swing-out rotor.
11. Resuspend vesicle pellets in 50 μl to 150 μl volume of TE buffer containing no sucrose (*see* **Note 29**).
12. Store membranes in liquid nitrogen until use.

3.3 Determination of Intactness and Purity of the Chloroplast and the Purity of the Envelope Fractions by Western Blotting

1. Take a 10 μl aliquot of the crude leaf homogenate, chloroplast, and/or chloroplast membrane fractions and mix them with 10 μl of 3-times concentrated PSB.
2. Incubate the samples at room temperature for 20 min.
3. Vortex the samples multiple times during the incubation and then add 10 μl DW.

4. To remove the insolubilized material from the samples, pellet the samples using 13,000 × *g* centrifugation at room temperature.
5. Collect the supernatant and add 2 μl bromophenol blue loading dye.
6. Perform SDS-PAGE on 10–18% gradient or 12% continuous polyacrylamide gels using a constant current of 20 mA per gel at 6 °C in case of 7 cm long, 1.5 mm thick gel system.
7. Using a protein standard of known protein concentration, determine the protein concentration in the samples of interest after SDS-PAGE using comparative densitometry [8] (*see* **Note 30**).
8. Load 10 to 20 μg solubilized protein in the wells, repeat Step 2, and perform the horizontal transfer of the separated proteins onto nitrocellulose membrane using 90 V constant voltage (<0.4 A) for 3 h at 4 °C.
9. Block the membranes using 3% (w/V) gelatin or 5% (w/V) non-fat milk powder in TBS, and subject them to the primary antibody treatment in TBS containing 1% (w/V) gelatin powder according to the specification of the antibody:
   (a) Chloroplast intactness: Leaf homogenate and chloroplast samples – anti-RbcL and anti-apoLHCII.
   (b) Chloroplast purity: Leaf homogenate and chloroplast samples – anti-AOX 1/2.
   (c) Membrane vesicle purity: IEP37, TPT, TOC75, AOX 1/2, apoLHCII.
10. Wash the membrane to remove the primary antibody: 2 times in TBST and 2 times in TBS for 15 min.
11. Subject the membrane to the secondary antibody treatment in TBS containing 1% (w/V) gelatin or non-fat milk powder.
12. Repeat Step 6. and perform the detection according to the applied secondary antibody.
13. To calculate the percentage of intactness of the chloroplast isolates, determine and compare the density of RbcL and apoLHCII bands in the chloroplast and the homogenate samples (*see* **Note 31**).
14. To check the purity of chloroplast envelope membrane fractions, apoLHCII, and AOX 1/2 contamination should be controlled in cIE and cOE samples. Purified cIE samples are also free of TOC75 signal (*see* **Note 32**).

3.4 Sample Preparation for Element Content Measurement by ICP-MS

1. Prepare fresh leaf material of approx. 1 g or a chloroplast suspension of approx. 500 μl volume in WB (isolated from 2–3 g leaf material).
2. To determine the element content in leaf samples, dry the leaf material at 60 °C for one week.
3. Measure the air-dry weight of the leaves on analytical scale.
4. After placing the leaf material for digestion in a max. 50 ml glass beaker (free of any organic and inorganic contaminations) add 2 ml of H_2O_2 and perform the digestion for 30 min at room temperature.
5. Mix the digested samples to ensure complete soaking.
6. Add 3 ml of cc. HNO_3 and cover the samples with watch glass. Place the samples on a hot plate placed in a laboratory fume hood and set at 60 °C and avoid boiling the samples. Samples should be mixed by gentle shaking every 15 min (*see* **Note 33**).
7. Continue the process to the complete dissolution of all remaining leaf material (the process takes 60–120 min, depending on the model taxon, and the physiological and developmental status of the leaves).
8. Cool down the samples in beakers to room temperature, rinse the convex side of the watch glass 2 times with 1–1 ml of deionized water into the beaker.
9. Collect the solution, filter it through MN 640 W paper, pour it into a volumetric flask, and fill it up to 10 ml with deionized water.
10. To determine the element content in chloroplasts, follow the procedure in Subheading 3.1, **steps 4–8** and **10** to prepare chloroplast samples for digestion. It is advised to use a concentrated suspension of chloroplasts in WB (i.e., approximately 2.5 million chloroplasts in 1 μl suspension, around 3 mg ml^{-1} total chlorophyll content) of 500–1000 μl volume. The quantification of chloroplast density, described in Subheading 3.1, **step 14**, is essential prior to element analysis.
11. Measure the samples in ICP-MS according to the operation instructions of the device.
12. Normalize the element concentrations of the samples calculated in ppb based on the ICP-MS analysis on a dry weight (leaf samples) or on a chloroplast density (chloroplast suspension samples) basis.

3.5 Iron Species Determination by Mössbauer Spectroscopy

1. Apply ^{57}Fe source grown or ^{57}Fe treated material for Mössbauer spectroscopy analysis of Fe species. To measure intact leaves, collect approx. 1 g of fresh leaf material, and freeze and store it in liquid nitrogen.

2. Pellet the chloroplast suspension by centrifugation as in Subheading 3.1, **step 8**. The total volume of the chloroplast pellet should be at least 5 ml. Chloroplast pellet should be frozen in liquid nitrogen and stored until measurement.
3. After the Mössbauer spectrometer is calibrated with α-iron at room temperature, measure the samples by a ^{57}Co(Rh) source of ~10^9 Bq activity in a liquid nitrogen bath cryostat at 80 K. To study the magnetic properties of Fe species, apply a Mössbauer spectrometer operating in conjunction with a cryomagnetic system generating 5 T magnetic field in a parallel geometry with respect to the propagation of the gamma radiation. Apply a cryostat operating at 5 K with liquid helium.
4. Calculate isomer shift (δ, mm s^{-1}), quadrupole splitting (Δ, mm s^{-1}), and partial resonant absorption areas (S_r, %) for the spectral components providing information regarding the electron densities at nuclei (valence state) and on the magnitude of any electric field gradients (coordination number).
5. Obtain quantitative analytical information of the species from the relative spectral areas [12].

3.6 Iron Uptake Measurement

1. Dilute the isolated class I chloroplasts by WB to 100 μg total chlorophyll ml^{-1}.

   Keep chloroplast suspension in darkness on ice constantly until all the components required for the Fe uptake experiment are added.

   Add the preferred chemical form of Fe to the chloroplast suspension according to the experimental setup.

   Mix the chloroplast suspension by gently turning it upside down.
2. To make references for the chloroplast Fe uptake determination, take 500 μl of aliquot out from the cooled suspension kept in darkness, in three replicates.
3. Warm up the chloroplast suspension to room temperature for 5 min prior to the start of the incubation. To prevent the sedimentation of the suspension, apply gentle but continuous shaking (e.g., with a swinging desk).
4. Initiate chloroplast Fe uptake by placing the chloroplast suspension at 120 μmol m^{-2} s^{-1} PPFD actinic light. If necessary, apply heat protection between the chloroplast suspension and the light source. For comparison, incubate a second aliquot of the chloroplast suspension under identical conditions except for the light irradiation (in darkness) (Fig. 1).
5. Terminate the Fe acquisition by placing samples onto ice in darkness and take 500 μl aliquots of the suspension in triplicates (*see* **Note 34**).

6. Wash out the chloroplasts from the uptake medium with centrifugation in a pre-cooled (4 °C) swing-out rotor at 2,500 × *g*, 5 min.
7. As for the removal of the surface attached Fe, resuspend the chloroplast pellet in 250 μl WB supplemented with 2 mM EDTA.
8. Repeat the centrifugation according to Step 7.
9. Resuspend the chloroplast pellet in 250 μl WB.
10. Add 50 μl 10% (w/V) SDS, 50 μl 10% (w/V) DTT, and 150 μl deionized water to the samples to solubilize chloroplasts. Incubate the samples at room temperature for 30 min.
11. Remove the non-solubilized material (mainly starch) from the samples by centrifugation at room temperature, 10,000 × *g* for 5 min in angle rotor.
12. Add 130 μl 1 mM ascorbic acid, 130 μl 3 mM BPDS, and 40 μl of deionized water to the sample to ensure the reduced status of total Fe contact for the formation of $[Fe(BPDS)_3]^{4-}$ complexes.
13. After 60 min incubation at room temperature in darkness, measure the absorbance of the samples at 535 nm using a UV-VIS spectrophotometer.
14. Calculate the Fe content of the samples using the absorption coefficient of $E = 22.14\ mM^{-1}\ cm^{-1}$ for $[Fe(BPDS)_3]^{4-}$ complexes [13].
15. Chloroplast Fe uptake is calculated on the basis of the difference in the Fe content of chloroplast suspensions samples prior to, and after the incubation. Normalize the Fe uptake based on the data of chloroplast density and intactness.

3.7 Chloroplast Envelope Ferric Chelate Reductase Activity Measurement

1. To measure chloroplast envelope membrane FCR (EC 1.16.1.9) activity, apply a chloroplast inner envelope vesicle suspension to 50 μg protein suspended in TE buffer.
2. To measure FCR reaction in 1.3 ml cuvette, add 10 (– 40) μl of 10 mM NADPH and 10 μl of 10 mM FAD to an aliquot (based on the protein concentration, approx. 10 μl) of the chloroplast inner envelope vesicle suspension (*see* **Note 35**).
3. Enclose the substances into the right-side-out membrane vesicles with a freeze-thaw cycle at –20 °C and 0 °C, respectively.
4. Add a minimum of 100 μl of 3 mM BPDS to the stock solution. BPDS forms complexes with ferrous Fe in 3:1 stoichiometry. Thus, at the application of high Fe concentration in the reaction mixture, the concentration of BPDS can be increased to avoid a BPDS limitation of the ferrous Fe detection.

5. Pipette the desired amount of Fe (III)-EDTA directly into the cuvette (*see* **Note 36**).
6. Transfer the reaction mixture to the cuvette and mix it with the ferric Fe component by diluting the reaction mixture with room temperature deionized water to achieve the final concentration of the components. Mind that this step initiates the FCR reaction.
7. Measure the changes in the absorbance at 535 nm in time using a spectrophotometer against a blank reaction that contains all components, except for the chloroplast inner envelope vesicle suspension (*see* **Note 37**). In the blank reaction, replace the volume of the envelope membrane samples with deionized water.
8. Measure the changes for 10–30 min to observe a clear linear increase in the kinetics.
9. Determine the amount reduced using the absorption coefficient of $E = 22.14\ mM^{-1}\ cm^{-1}$ for $[Fe(BPDS)_3]^{4-}$ complexes [13].

3.7.1 Ferric Chelate Reductase Activity Measurement on Solubilized Chloroplast Inner Envelope Samples (See Note 38)

1. To measure the enzyme activity in 1.3 ml cuvettes, add 26 μl of 1% (V/V) Triton X-100 to achieve a final concentration of 0.02% (V/V) Triton X-100 to a chloroplast inner envelope vesicle suspension of to 50 μg protein suspended in TE buffer. Solubilize the membrane vesicles for 30 min on ice.
2. Add 10 (– 40) μl of 10 mM NADPH, 10 μl of 10 mM FAD, and a minimum of 100 μl of 3 mM BPDS stock solutions to the solubilized mixture.
3. Apply Subheading 3.7, **step 5**, **6**, and 7. to measure FCR activity.

3.8 Transient Expression to Test Chloroplast Targeting of Iron Homeostasis Elements

1. Synthesize or conduct a PCR to amplify the cDNA of the gene of interest and clone it into a constitutive plant expression vector in frame with the GFP tag.
2. Transform the construct into competent AGL1 cells.
3. Grow the Agrobacteria for 2 days at 28 °C in LB broth supplemented with the respective antibiotics using a rotary shaker set at 200 rpm.
4. Collect the cells by pelleting at 2,000 × *g* for 2 min by centrifugation. Suspend the cells in Agro-infiltration medium to reach OD_{600} 0.6.
5. Supplement the suspension with 150 μM acetosyringone and keep it agitated at room temperature for 2–3 h prior to the infiltration [14].

6. Using a needle-less 1 ml disposable syringe, inject the suspension into the abaxial surface of the leaves with gentle pressure.
7. GFP fluorescence can be detected by a handheld UV or blue lamp and orange filter in 2–4 days after the infiltration.
8. Cut the leaf parts that show GFP fluorescence, put it on ice, and continue to isolate protoplast (according to Subheading 3.9) or chloroplast (according to Subheading 3.1) from the leaf pieces.

3.9 Protoplast Isolation

1. Isolate protoplasts from both the infiltrated and wild-type leaves to study the autofluorescence. Cut the leaves with a fresh blade into 1–2 mm strips and immediately immerse them into PWB to avoid them drying.
2. Replace the PWB with PIB, and vacuum infiltrate it for 10 min. The recommended leaf buffer ratio is 50–100 mg leaf material in 1 ml solution.
3. Cover the beaker with parafilm to avoid evaporation of the buffer. Incubate at room temperature in the dark for 3–16 h with gentle shaking (30–70 rpm) or without agitation. The optimal digestion parameters may differ from species to species (*see* **Note 39**).
4. Take 20 μl samples and observe the quality and quantity of the protoplasts in light microscopy during the incubation.
5. When the released protoplast number reaches the optimum levels, filter them through 2–3 layers of gauze, then filter again on the 100 μm cell strainer.

3.10 Localization of the Transiently Expressed GFP Fused Protein

1. Dye the protoplasts with the dyes that should be used before fixation (e.g., BioxMLYellow (optional)).
2. Mix 50–100 μl the protoplast suspension with 2-times concentrated PFM in 1:1 ratio.
3. Perform the fixation for 20 min in darkness.
4. Add 1 ml IWB, incubate with it for 5 min and pellet the protoplasts by centrifugation at 150 × *g* for 3 min in swing-out rotor.
5. Repeat the washing step again. Continue with:
   (a) Immunofluorescent labeling (*see* Subheading 3.10, **step 3**.) or.
   (b) Stain using DAPI (optional) then mount them to microscope slide.
6. Block and permeabilize the protoplasts for 20 min in IBB then subject them to the primary treatment in PBS containing 1% (w/V) BSA according to the specification of the anti-GFP antibody.

7. Wash the protoplasts 2 times in IWB for 5 minutes and centrifuge them at 150 × *g* for 3 min.
8. Subject the protoplasts to the fluorescently labeled secondary antibody, repeat the washing steps 2 times, and continue with Subheading 3.10, **step 5b**.
9. Observe the mounted protoplasts in confocal fluorescence microscopy using.
   (a) UV laser to excite DAPI; detection between 415–460 nm,
   (b) 488 nm laser to excite GFP; detection between 498–530 nm,
   (c) 543 nm laser to excite BioxMLYellow or secondary antibodies conjugated to Atto 550; detection between 553–595 nm,
   (d) 638 nm laser to excite the chlorophylls with detection between 645–790 nm.

4 Notes

1. Abiotic stressors as well as the light intensity have a high impact on the accumulation of carotenoids on the chloroplast inner envelope membrane. Thus, a controlled environment is recommended to achieve a stable presence of carotenoids in the chloroplast inner envelope membrane of the model system.
2. The requirement of plants for nutrients is taxon-specific. In our system, quarter-strength and half-strength Hoagland's solutions proved to be optimum for sugar beet and oilseed rape models. It is important to mention that the sugar beet model requires the addition of 0.25 mM NaCl in the nutrient solution in order to avoid root neck rotting. In contrast to the sugar beet model, the oilseed rape model has a higher dose requirement for nutrients, and thus it is recommended to cultivate this model on a half-strength medium.
3. The chemical form and amount of Fe in the nutrient medium should be chosen in accordance with the aim of the experimental setup. In case a pre-cultivation is required for the applied plant model to achieve a proper size and physiological status for Fe deficiency treatment, only Fe (III)-EDTA (or other strong Fe (III) chelates) can be applied during the pre-cultivation period. In contrast, the application of EDTA can modify the effect of certain heavy metal treatments. Thus, in such an experimental setup, the application of natural Fe (III) complexes such as Fe (III)-citrate is recommended. To grow plant material for ^{57}Fe Mössbauer spectroscopy analysis, it is highly important to cultivate the plant material on an Fe source that is prepared from non-radioactive ^{57}Fe enriched Fe source.

4. For the functional analysis of chloroplast Fe uptake and homeostasis, most techniques require a large amount of plant material to isolate and purify chloroplasts and chloroplast envelope membranes. Although spinach is often a subject of chloroplast studies, its hydroponic cultivation is problematic. Sugar beet is a close relative and thus a reasonable alternative of spinach in chloroplast Fe homeostasis studies. Oilseed rape also produces a large amount of leaf material, and its model is also easy to cultivate by hydroponics. Since Oilseed rape is a close relative to *Arabidopsis thaliana*, it is a reasonable alternative of this latter model in functional analysis studies that require a large amount of plant material.
5. If necessary, IB free of sodium ascorbate can be stored at 4 °C for maximum a week; sodium ascorbate should be added just before use. Keep in mind that the presence of EDTA can corrupt the later Fe uptake experimental setup.
6. In case of proceeding to chloroplast envelope membrane analysis, 2 mM EDTA can be applied.
7. Digitalization of polyacrylamide protein gels can be performed, among others, by gel scanners.
8. Since is essential to exclude mitochondrial contamination, make sure to include antibodies against abundant mitochondrial proteins (such as AOX 1/2).
9. Although the chemical form of intracellular Fe that is subject of chloroplast Fe acquisition (also called as the labile Fe pool) has not been fully understood yet, chloroplasts prefer the utilization of Fe (III)-citrate 1:1 stoichiometry complex in the Fe uptake. Thus, if the chloroplast Fe uptake study does not aim to utilize a specific Fe compound, the application of Fe (III)-citrate 1:1 stoichiometry complex is suggested.
10. Chloroplast suspensions are sensitive against heat; thus, heat production during illumination should be limited. Alternatively, heat protection can be also achieved by placing chloroplast suspension under a Petri dish filled with water of 10 mM $CuSO_4$ during illumination.
11. EDTA may damage membranes; thus, keeping its concentration precisely is highly important.
12. To achieve a proper ferrous Fe complexing of BPDS in the solubilized chloroplast suspension, ascorbic acid, and BPDS are only added once the chloroplast suspension is already solubilized. Ascorbic acid stock solution should be prepared freshly. BPDS is light sensitive, especially in solution, thus it should be also prepared freshly and protected from light. Light exposure of BPDS solution induces false coloration that corrupt the

proper Fe determination. Ferrous Fe complexing of BPDS is sensitive to pH thus a quantitative measurement is only possible in a buffered medium.

13. Triton X-100 is a non-ionic detergent that destructs membrane structure without having a significant impact on activity (and thus on the structure) of proteins. Since Triton X-100 is a highly viscous fluid, it can be also added on m/V basis. The concentration of 0.02% (V/V) in the solubilization is not a strict value. Nevertheless, modification of the Triton X-100 concentration requires approval through unchanged FCR activity.
14. For *Phaseolus vulgaris* L., any other model can be applied in transient expression studies. Mind that the transient study should be performed on plants during the adequate developmental and physiological stage, considering the aim of the study (e.g., Fe storage-related proteins in plants grown on excess amounts of Fe).
15. The additive-free PB should be heated up to 40 °C to aid the solubilization of the protoplast digestion enzymes. After complete dissolution of the enzymes, cool the solution to room temperature and add the rest of the additives.
16. BSA reduces surface tension and aids pelleting the fixed and permeabilized cells.
17. To prepare the protoplast fixative, methanol-free 16% (w/V) PFA will be mixed with 2-times concentrated protoplast fixation buffer in 1:1 ratio.
18. GFP fluorescence imaging may interfere with autofluorescence emerging from the secondary metabolites of the plant model. Thus, the autofluorescence spectrum of the model taxon of identical physiological stage should be screened. In case autofluorescence emerges where GFP is also detected (*see* Subheading 3.10, **step** 7), a fluorescent secondary antibody of a non-overlapping emission peak should be used to prove the cellular localization of the GFP fusion protein. Yellow fluorescent antibodies (e.g., $Atto_{550}$) with green excitation do not usually interfere with the autofluorescence of plant compounds, neither with that of GFP.
19. Membrane dyes aid the visualization of the chloroplast envelopes. Do not use them in the continuation to immunofluorescence labeling due to the permeabilization.
20. To protect the blade of the blender, remove the main vein of the leaves. Depending on the type and operation of the blender (speed and breaking scheme) maximum 3 times, 1–1 s blending works better. The primary aim of the application of a blender is to mechanically break up as many mesophyll cells

as possible. Shear force, indeed, can mechanically damage the liberated plastids. Therefore, avoid a long, high-speed homogenizing.

21. Based on the model taxon and the physiological and developmental status, slight differences are applied in the centrifugation scheme. Chloroplasts originate from stressed plants that are less dense; thus, an effective pelleting requires higher speed.
22. All centrifugation steps should be carried out in swing-out rotor to minimize damaging the chloroplasts at pelleting. In an angle rotor, chloroplasts get pelleted on the side wall of a centrifuge tube and roll down to the tube bottom. That always results in a higher proportion of mechanically damaged chloroplasts.
23. As the pellet of intact chloroplasts is stable, the pellet does not flow down when gently removing the supernatant, the pellet does not flow.
24. Centrifugation results in the pelleting of the liberated starch granules. Intact (class I) chloroplasts remain on the 45/60% (w/V) sucrose interphase, while broken (class II) chloroplasts accumulate on the 20/45% (w/V) sucrose interphase (Fig. 2a).
25. To avoid damages at pelleting, a small amount (less than 1 ml in a 50 ml centrifugation tube) of 60% (w/V) sucrose in GB can be applied on the bottom of the tube during centrifugation.
26. The application of Bürker chamber is not compatible with 100× immersion objectives. At performing microscopy studies on an isolated chloroplast suspension, it is important to mind that isolated chloroplasts are sensitive objects that can be damaged by the mechanical pressure of the microscopy cover glasses. Moreover, in a chloroplast suspension the isolated chloroplasts sediment by their density. Thus, to achieve a microscopy image with all chloroplasts in focus, a wait time of approx. 5 min time is recommended prior to the microscopy studies. Chloroplasts are also sensitive to both heat and illumination, so a chloroplast suspension on a microscopy plate is also suggested to keep in darkness, on ice.
27. One-hour incubation time is crucial to reach the high purity of cIE membranes (i.e., reduced thylakoid cross-contamination).
28. Despite intact chloroplast isolation, separation of thylakoids results in a pellet that slightly flows- thus, at collecting the supernatant, take care to avoid thylakoid contamination. A dense pellet might indicate that the fracture of chloroplasts and removal of envelope membranes is incomplete/failed.
29. This usually provides a protein concentration of 4–8 μg μl^{-1} in the membrane suspension, with starting material of approx.

300 g leaves, resulting in 10 ml chloroplast suspension, used to generate 30 ml envelope vesicle enriched suspension.

30. Since the Bradford assay-based protein concentration estimation methods are not trustworthy on membrane proteins, we suggest determining the protein content of the samples using a protein gel electrophoresis, and density comparison-based techniques as performed in [8]. Although this protein concentration estimation technique has a significantly higher time commitment, the secondary advantage of the technique is that it does not just information on the protein concentration, but also on the quality of the protein within.

31. Intactness of the chloroplast isolates is acceptable if the ratio of the soluble RbcL to thylakoid apoLHCII in the chloroplast samples is higher than 80% of that in the crude leaf homogenate. Decreased proportion of RbcL indicates its removal during the washing steps through membrane damages.

32. Chloroplast inner envelope membrane vesicle fractions can be contaminated by thylakoids and outer envelope membrane vesicles. In cases of preparing vesicles for chloroplast inner envelope FCR activity measurement, traces of thylakoid contamination should be excluded. Illuminated thylakoids can catalyze Hill reaction—the reduction of ferric Fe by the activity of photosystem I reaction centers.

33. The cc. HNO_3 is highly corrosive thus it requires careful handling according to the proper safety precautions. Since cc. HNO_3 is also sensitive to light and heat, it should be kept in a protected place. Discoloration of cc. HNO_3 indicates the decomposing initiated by light. Also, be careful with the hot glassware when mixing the samples.

34. Fe uptake of chloroplasts is dependent on light and temperature and can be terminated by placing them in darkness. Nevertheless, a rapid decrease in temperature is required. Thus, it is recommended to pre-cool microcentrifuge tubes and collect the aliquots therein. Taking small quantity aliquots, such as 500 μl samples also supports the rapid cooling of the chloroplast suspension. It is recommended to work with triplicates.

35. The optimum final NADPH concentration in the reaction mixture is 100 μM NADPH. Nevertheless, to avoid NADPH limitation in the reaction, a higher NADPH concentration may be applied- but keep in mind that NADPH to ferric Fe background reaction is dependent on the NADPH concentration.

36. Applying Fe (III)-EDTA is suggested because of the rapid background reaction of NADPH, and weakening ferric Fe complexes may prevent the measurement of the enzyme activity in time.

37. Since Fe (III)-EDTA and BPDS are absent in the vesicles, the FCR reaction can only be mediated by right-side-out vesicles.
38. Taking no consideration of the orientation of the proteins in the chloroplast inner envelope membrane, FCR activity can be also measured on gently solubilized chloroplast inner envelope membrane vesicle samples in cases where gentle solubilization does not affect the enzyme activity, but liberates proteins from the membranes.
39. Protoplast isolation from different taxa and/or grown under different conditions (e.g., Fe deficiency, heavy metal stress, heat stress) may require slight modifications of the isolation parameters. In the fine-tuning of the isolation protocol, both macro and microscopic observations could help. In order to resolve the problem of browning of the suspension, add PVPP or ascorbic acid to avoid oxidation. If cells appear in the isolate with the remaining cell wall, increase the amount of cellulase. To minimize broken or disrupted cells, decrease the digestion time and/or avoid shaking.

Acknowledgments

The Authors acknowledge the help of Barnabás Cseh, Brigitta Müller, Máté Sági-Kazár, and Éva Sárvári for discussing the protocol details and for their useful comments. The Authors also acknowledge the financial support from the National Research, Development and Innovation Office, Hungary (NKFIH K-135607).

References

1. Hantzis LJ, Kroh GE, Jahn CE et al (2018) A program for iron economy during deficiency targets specific Fe proteins. Plant Physiol 176: 596–610. https://doi.org/10.1104/pp.17.01497
2. Higuchi K, Saito A (2022) Elucidation of efficient photosynthesis in plants with limited iron. Soil Sci Plant 68:505. https://doi.org/10.1080/00380768.2022.2106115
3. Balk J, Pilon M (2011) Ancient and essential: the assembly of iron–sulfur clusters in plants. Trends Plant Sci 16:218–226. https://doi.org/10.1016/j.tplants.2010.12.006
4. Vigani G, Murgia I (2018) Iron-requiring enzymes in the spotlight of oxygen. Trend Plant Sci 23:874–882. https://doi.org/10.1016/j.tplants.2018.07.005
5. Jones JD (2020) Iron availability and management considerations: a 4R approach. Crop Soil 53:32–37. https://doi.org/10.1002/crso.20019
6. Sági-Kazár M, Solymosi K, Solti Á (2022) Iron in leaves: chemical forms, signalling, and in-cell distribution. J Exp Bot 73:1717–1734. https://doi.org/10.1093/jxb/erac030
7. Hill R, Scarisbrick R (1940) The reduction of ferric oxalate by isolated chloroplasts. Proc Royal Soc London Ser B-Biol Sci 129:238–255. https://doi.org/10.1098/rspb.1940.0038
8. Sárvári É, Gellén G, Sági-Kazár M et al (2022) Qualitative and quantitative evaluation of thylakoid complexes separated by blue native PAGE. Plant Method 18:1–19. https://doi.org/10.1186/s13007-022-00858-2
9. Porra RJ, Thompson WA, Kriedman PE (1989) Determination of accurate excitation coefficient and simultaneous equations for

assaying chlorophylls *a* and *b* extracted with four different solvents: verification of the concentration of chlorophyll standards by atomic absorption spectroscopy. Biochim Biophys Acta 975:384–394. https://doi.org/10.1016/S0005-2728(89)80347-0

10. Keegstra K, Yousif AE (1986) Isolation and characterization of chloroplast envelope membranes. Method Enzymol 118:316–325. https://doi.org/10.1016/0076-6879(86)18082-7
11. Froehlich JE, Wilkerson CG, Ray WK et al (2003) Proteomic study of the *Arabidopsis thaliana* chloroplastic envelope membrane utilizing alternatives to traditional two-dimensional electrophoresis. J Proteome Res 2:413–425. https://doi.org/10.1021/pr034025j
12. Greenwood NN, Gibb TC (1971) Mössbauer spectroscopy. Chapman and Hall, London
13. Smith GF, McCurdy WH, Diehl H (1952) The colorimetric determination of iron in raw and treated municipal water supplies by use of 4:7-diphenyl-1:10-phenanthroline. Analyst 77:418–422. https://doi.org/10.1039/AN9527700418
14. Zhang Y, Chen M, Siemiatkowska B et al (2020) A highly efficient agrobacterium-mediated method for transient gene expression and functional studies in multiple plant species. Plant Commun 1:100028. https://doi.org/10.1016/j.xplc.2020.100028

Chapter 13

Perls/DAB Staining to Examine Iron Distribution in Arabidopsis Embryos

Joaquín Vargas and Hannetz Roschzttardtz

Abstract

Iron is accumulated in Arabidopsis embryos during seed maturation. Where iron localizes in seed and embryo is important information for seed research. Iron detection can be performed in an inexpensive manner using Perls staining, based on the Prussian blue complex formation. After this first step, DAB intensification can be performed in order to visualize easily where iron pools are located in isolated embryos.

Key words Iron staining, Embryo, Perls staining, Prussian Blue, DAB intensification

1 Introduction

Iron is an essential micronutrient. During seed development, iron accumulates in embryo cells [1]. The analysis of iron distribution at the cell and subcellular level using histological methods has allowed the ability to describe the idea that plants do not have the same strategy to locate iron in embryos. For instance, *Arabidopsis thaliana* embryos accumulate iron in vacuoles of only one cell layer [1], in contrast, *Chenopodium quinoa* accumulates iron in several cell layers, including cortex and endodermis/bundle sheath [2]. These results were obtained by Perls/DAB staining, an easy and non-expensive approach that can be used to reveal iron localization in whole embryos and on thin sections [1, 3]. Iron localization obtained by Perls/DAB has been corroborated by analytical methods like micro X-ray fluorescence (μXRF), validating its specificity [2, 4, 5]. In this protocol, we describe how to perform Perls/DAB staining on isolated Arabidopsis embryos.

Jeeyon Jeong (ed.), *Plant Iron Homeostasis: Methods and Protocols*, Methods in Molecular Biology, vol. 2665,
https://doi.org/10.1007/978-1-0716-3183-6_13,

2 Materials

Prepare all solutions using Milli-Q water and analytical-grade reagents.

2.1 Seed Dissection/Embryo Isolation

1. Stereo microscope
2. Petri plate
3. Twitters
4. 1.5 mL plastic tube

2.2 Perls Staining

1. 4% HCl: Add 1,08 mL of 37% HCl (concentrated 12 N HCl) to 8,92 mL of water (*See* **Note 1**).
2. 4% Potassium Ferrocyanide: Weigh 2 g of Potassium Ferrocyanide and prepare 10 mL of solution (*See* **Note 2**).

2.3 DAB Intensification

1. 0.1 M phosphate buffer (pH 7.4): Stock Solution A (0.2 M Sodium Phosphate Monobasic): Weigh 27.6 g of NaH_2PO4*H_2O (FW: 137.99 g/mol) and prepare 1 L of solution with H_2O. Stock Solution B (0.2 M Sodium Phosphate Dibasic): 53.62 g $Na_2HPO4*7H_2O$ (FW: 268.07 g/mol) and prepare 1 L of solution. To prepare 100 mL of 0.1 M phosphate buffer (pH 7.4), mix 10 mL of solution A and 40 mL of solution B, and add H_2O up to 100 mL.
2. Methanol/NaN_3/H_2O_2: Weigh 0,13 g of NaN_3 (FW: 65 g/mol) and add 198 mL of methanol. Add 2 mL of 30% (v/v) H_2O_2. (*See* **Notes 1** and **2**).
3. 1% $CoCl_2*6H_2O$: Weigh 1 g of $CoCl_2*6H_2O$ and prepare 100 mL of dissolution with H_2O.
4. DAB-HCl/H_2O_2/$CoCl_2$ Solution: Weigh 25 mg of 3,3′-diaminobenzidine-4HCl and add 0.1 M phosphate buffer (pH 7.4) to complete 99,5 mL. Add 17 μL of 30% (v/v) H_2O_2 and 500 μL of 1% $CoCl_2*6H_2O$. (*See* **Notes 1** and **2**).

3 Methods

3.1 Seed Dissection/Embryo Isolation

1. When dry seeds are used, hydrate them by incubation in Milli-Q water for at least 3 h at 4 °C.
2. Using a stereo microscope, tweezer, and a needle, carefully peel the seeds in order to remove integuments and liberate the embryo. During this step, seeds and isolated embryos must be submerged all the time to prevent dehydration.

3. Once a number of embryos are obtained, transfer them into a 1.5 mL plastic tube using a Pasteur pipette. Remember, embryos must be submerged all the time.

3.2 Perls Staining

1. Carefully remove the water from the 1,5 mL tube containing the embryos using a P200 micropipette.
2. Add 500 μL of 4% HCl and 4% Potassium ferrocyanide (final concentration: 2% of HCl and Potassium Ferrocyanide)
3. Vacuum for 45 min. As a control, incubate isolated embryos in 2% of HCl without Potassium Ferrocyanide.
4. After vacuum, remove the solutions and wash the embryos three times using 1 mL of Milli-Q water.
5. Potassium ferrocyanide and iron accumulated in the embryos will form a blue complex called Prussian Blue, that can be observed directly in a stereo microscope. A photographic recording is recommended. In case a DAB intensification is needed, follow the method in Subheading 3.3.

3.3 DAB Intensification

1. Incubate Perls-stained and control embryos with 1 mL of Methanol/NaN_3/H_2O_2 solution for 1 h.
2. Carefully remove the Methanol/NaN_3/H_2O_2 solution using a P200 micropipette and wash the embryos two times with 1 mL of Milli-Q water.
3. Incubate the embryos with 0.1 M phosphate buffer (pH 7.4) for 5 min.
4. The final step of intensification is done by incubating embryos in DAB-HCl/H_2O_2/$CoCl_2$ solution for 10 min.
5. Remove DAB-HCl/H_2O_2/$CoCl_2$ solution and wash the embryos three times with 1 mL of Milli-Q water.
6. Keep the embryos in Milli-Q water during observation and photographic recording.

4 Notes

1. All operations involving concentrated acids and methanol must be conducted in the fume hood and the investigator should wear the appropriate personal protective equipment.
2. Prepare fresh solutions.

Acknowledgments

This work was supported by a grant from FONDECYT-Chile 1160334 to HR, Facultad de Ciencias Biológicas-PUC grant to HR, and CONICT-Chile grant 21151344 to JVP.

References

1. Roschzttardtz H, Conejero G, Curie C, Mari S (2009) Identification of the endodermal vacuole as the iron storage compartment in the Arabidopsis embryo. Plant Physiol 151:1329–1338
2. Ibeas M, Grant-Grant S, Coronas M, Vargas-Pérez J, Navarro N, Abreu I, Castillo-Michel H, Avalos-Cembrano N, Paez-Valencia J, Pérez F, González-Guerrero M, Roschzttardtz H (2019) The diverse iron distribution in Eudicotyledoneae seeds: from Arabidopsis to Quinoa. Front Plant Sci 15:1985
3. Meguro R, Asano Y, Odagiri S, Li C, Iwatsuki H, Shoumura K (2007) Nonheme-iron histochemistry for light and electron microscopy: a historical, theoretical and technical review. Arch Histol Cytol 70:1–19
4. Kim S, Punshon T, Lanzirotti A, Li L, Alonso J, Ecker J, Kaplan J, Guerinot M (2006) Localization of iron in Arabidopsis seed requires the vacuolar membrane transporter VIT1. Science 314:1295–1298
5. Roschzttardtz H, Grillet L, Isaure MP, Conéjéro G, Ortega R, Curie C, Mari S (2011) Plant cell nucleolus as a hot spot for iron. J Biol Chem 286:27863–27866

Chapter 14

Visualizing Metal Distribution in Plants Using Synchrotron X-Ray Fluorescence Microscopy Techniques

Ju-Chen Chia, Arthur R. Woll, Louisa Smieska, and Olena K. Vatamaniuk

Abstract

Recent improvements in synchrotron-based X-ray fluorescence (SXRF) microscopy established it as an advanced analytical tool for analyzing 2D- and 3D distribution of mineral elements in plants. Among existing imaging techniques, SXRF microscopy offers several unique capabilities, including in situ metal quantification in plant tissues and high sensitivity, as low as 1 mg kg^{-1}, at the nanoscale spatial resolution. SXRF is increasingly utilized in different plant science disciplines to provide a fundamental understanding of metal homeostasis, and the function of trace elements in plant metabolism and development. Here, we describe methods for SXRF imaging, including sample preparation, the optimization of conventional SXRF for analyzing trace elements, and the development of confocal SXRF (C-SXRF).

Key words Synchrotron-based X-ray fluorescence microscopy, Confocal, Plants, Metallomics, Iron, Copper, Spatial distribution, In situ analysis

1 Introduction

The development of in situ chemical analytical tools for visualizing trace elements distribution in plant tissues can improve our understanding of plant metabolism and development, and assist in assigning functions to unknown genes. Methods suitable for examining and quantifying the concentration and distribution of total element pools within plants include synchrotron-based X-ray fluorescence (SXRF) microscopy, particle-induced X-ray emission (PIXE), scanning or transmission electron microscopy coupled with energy dispersive X-ray spectroscopy (SEM-EDXS or TEM-EDX), laser ablation inductively coupled plasma mass spectrometry (LA-ICP-MS), and nanoscale secondary ion mass spectrometry (SIMS) [1–3]. While the first three approaches are X-ray fluorescence (XRF)-based technologies, the latter two are mass

Ju-Chen Chia and Arthur R. Woll contributed equally.

Jeeyon Jeong (ed.), *Plant Iron Homeostasis: Methods and Protocols*, Methods in Molecular Biology, vol. 2665, https://doi.org/10.1007/978-1-0716-3183-6_14,

spectrometry-based methods. These existing imaging techniques offer resolutions ranging from micron to submicron scales, and only EDXS and XRF microscopy provide a non-destructive analysis [1, 2]. Because EDXS systems usually operate under a high vacuum and require dehydrated specimen unless a cryo-system is available, in vivo analyses of element distribution using EDXS, especially at room temperature, are typically challenging [2]. SXRF techniques, on the other hand, allow in situ quantification analysis and high element sensitivity at 1 mg kg^{-1} [1, 2, 4, 5]. Other non-destructive and in situ analysis methods include histochemical detection with chromogenic and fluorogenic dyes or autoradiography with radioactive isotopes [1–3, 6]. These techniques are semi-quantitative and their capability for in situ metal distribution studies is either limited by the availability of fluorophores with specific affinities for elements of interest, or in the case of tracer studies, cannot provide direct information on the endogenous metal composition of cells or tissues and mainly provide information on metal uptake, distribution and compartmentation [1–3, 6].

Recent advances in SXRF imaging including high-speed X-ray detectors, confocal X-ray optics, and high-resolution scan-probe imaging, now provide rapid, non-destructive, two- and three-dimensional images of entire plants with up to sub-micron resolution. These images provide direct access to the plant's structure, the spatial distribution of metals and other compounds, and their chemical state. Coupled with the ability of modern genetic tools, such as CRISPR/Cas9, to manipulate genetic information, these X-ray data provide direct information about the function of the specific enzymes and proteins encoded by the target genes, increasing our knowledge of the genotype-to-phenotype relationship, simultaneously advancing our understanding of the mechanisms governing life and providing the basis for engineering safe and effective new crop varieties [4, 7–9]. Here, we describe different procedures for the preparation of plant samples for conventional SXRF and the optimization of SXRF methods for the in situ visualization and quantification of low-abundance metal ions (e.g., copper). We also present a method for trace elements visualization using *high-resolution* confocal SXRF (C-SXRF) microscopy, established at the Cornell High Energy Synchrotron Source (CHESS). The advantage of C-SXRF over conventional SXRF is to provide direct access to a fixed 3D volume in space. High-resolution C-SXRF, which relies on Collimating Channel Arrays (CCAs) (invented and fabricated by CHESS), provides spatial resolution reaching 2 μm from 2–20 keV, as much as 20× better than other implementations of C-SXRF microscopy. C-SXRF has been already successfully used for analyzing the distribution of copper, iron, manganese, and zinc in the vascular tissues from the leaf blade and the node sections of *Arabidopsis thaliana* and *Brachypodium distachyon*, respectively [10, 11].

2 Materials

2.1 Mounting Materials

1. Kapton film, 0.3 mil thickness.
2. Kapton tape, 1 mil thickness (*see* **Note 1**).
3. Sample holder (*see* **Note 2**).
4. Plastic replaceable tip tweezers.
5. Kimwipes.
6. 95% ethanol.

2.2 Sample Fixation and Embedding

1. Fixation solution: 2% glutaraldehyde in 0.05 M cacodylate buffer pH 7.4. The solution needs to be prepared freshly.
2. Absolute ethanol to make an ethanol series of 25%, 50%, 70%, 95%, and 100% ethanol.
3. Acetone.
4. Molecular sieves, 3A.
5. Spurr's low-viscosity embedding media kit (Electron Microscopy Science, Catalog #14300). Prepare the embedding media according to the manufacturer's instructions.

2.3 Conventional 2D Scanning SXRF Equipment

Conventional 2D scanning SXRF has been performed at CHESS beamlines F3 and ID3B/FMB.

Essential beamline characteristics and equipment include:

1. Monochromatic focused X-rays. Example energies and focal spot sizes include:
   (a) CHESS F3: 17 keV with a focusing mono capillary yielding a roughly 25 μm focused beam containing $\sim 1^{10}$ photons/s.
   (b) CHESS ID3B/FMB: 9.7 keV with beryllium compound refractive lenses yielding a roughly 2 μm tall × 12 μm wide focused beam containing $>3^{10}$ photons/s.
2. X-ray fluorescence detector: for example, a Vortex ME4.
3. An aligned X-ray fluorescence setup (provided by the beamline) comprising a focused beam, motorized stages, a digital signal processor (e.g., Quantum Xspress3), and hardware to support fly-scanning. A schematic is shown in Fig. 1.
4. A set of XRF calibration foils, for example, purchased from Micromatter Technologies, Inc., consisting of known areal concentrations between 10 and 20 μg/cm^2 deposited on 6.3 μm Mylar film. The most frequently used foils are iron and copper.

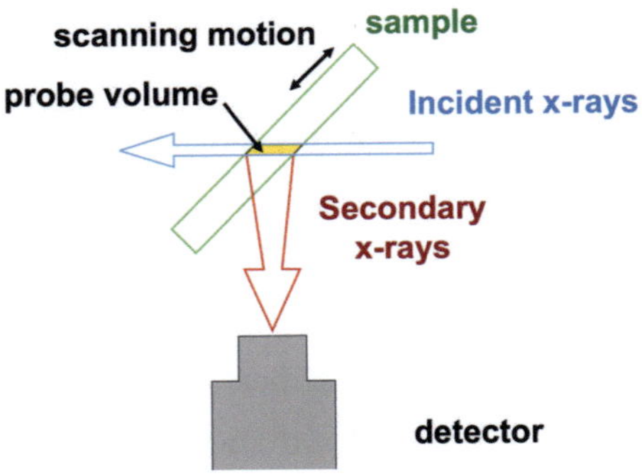

Fig. 1 Schematic of typical 2D scanning XRF experiment

5. Beamline hardware & control software to support scanning and data acquisition.
6. Software for data processing (e.g., Praxes).

2.4 Confocal SXRF Equipment

Confocal SXRF (C-SXRF) was performed at the NSLS-II beamline 5-ID/SRX and used to determine volumetric concentrations of the elements like potassium, calcium, manganese, iron, copper, and zinc in biological tissues.

Essential beamline characteristics & equipment:

1. Monochromatic beam with X-ray energy between the energy of 10.0 keV, focused to approximately 2 × 2 microns^2 and flux of approximately 6×10^{11} photons/second in the focal spot.
2. A single-element Vortex EX Energy dispersive detector.
3. A collimating channel array (CCA) comprising 175, 2-micron channels etched into a germanium substrate. The channels are all directed to a common intersection, defining the probe volume. The channels are etched to a depth of approximately 300 microns. The CCA device has the shape of a trapezoid and is mounted into a custom mount, which provides shielding from unwanted X-ray background, and which mounts to the single-element vortex detector.
4. An aligned C-SXRF setup employing focusing optics for the incident beam (provided by the beamline) and the CCA described above. The setup is shown schematically in Fig. 2.
5. A thin-film XRF reference sample purchased from Applied X-ray Optics (Dresden GmbH); Item number RF17–200-S4216–02 comprising a glassy layer on a thin silicon nitride window. The glassy layer contains six metals at known areal mass densities, namely lead, lanthanum, palladium, molybdenum, copper, and iron at concentrations of 8.49+/−1.23, 12.14 +/− 1.45, 2.33 +/− 0.45, 0.86 +/− 0.09, 2.23 + 0.33, and 4.39 +/− 0.59 μg/cm^2.

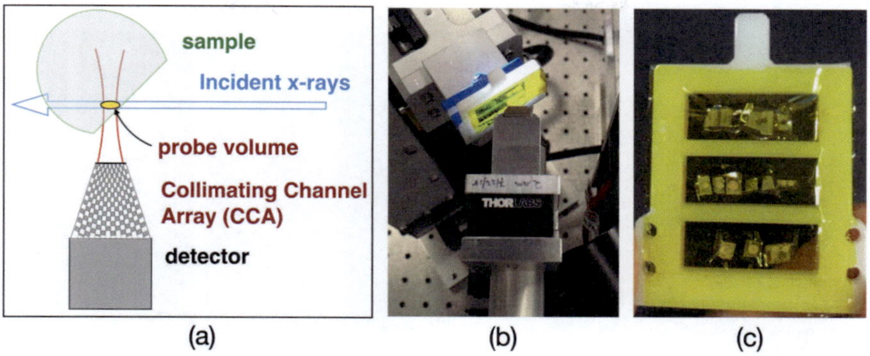

Fig. 2 Schematic of C-SXRF experiment. (**a**) schematic diagram of C-SXRF experimental setup employing CCAs as the collection optic. (**b**) photograph of the setup as used at NSLS-II beamline SRX/5-ID. (**c**) photograph of a sample holder with a variety of samples mounted. This mounting scheme constrains the front surface of the samples to approximately the same plane in the laboratory frame

6. Kinematic sample mount at the scan position for easy, reproducible placement of samples in the scan position.
7. Alignment microscope facing the sample surface with ~0.1 mm field of view, ~0.01 depth of focus.
8. Beamline hardware & control software to support scanning and data acquisition.

3 Methods

3.1 Sample Preparation and Mounting

3.1.1 Fresh Samples

Samples Mounted in a Wet Chamber

1. Seal one side of the sample holder with Kapton film.
2. On a clean surface, place a piece of Kapton tape with the sticky end facing up.
3. Detach the samples from the plants and arrange them on the Kapton tape immediately.
4. Place the samples along with the tape on the Kapton film on the sample holder.
5. Wet a piece of Kimwipe with deionized water and make a roll.
6. Place the wet Kimwipe at the bottom of the sample holder to keep the fresh sample hydrated.
7. Seal the sample holder with another layer of Kapton film and analyze the element distribution immediately (*see* **Note 3**). Figure 3 shows the setup of a wet chamber.

Live Plant Tissues

If feasible, mounting the fresh tissues without detaching them from the plants could better prevent dehydration of the samples during SXRF scans. This method requires an area next to the sample station to place plants in pots, or a shelf on the sample translation stage to hold pots.

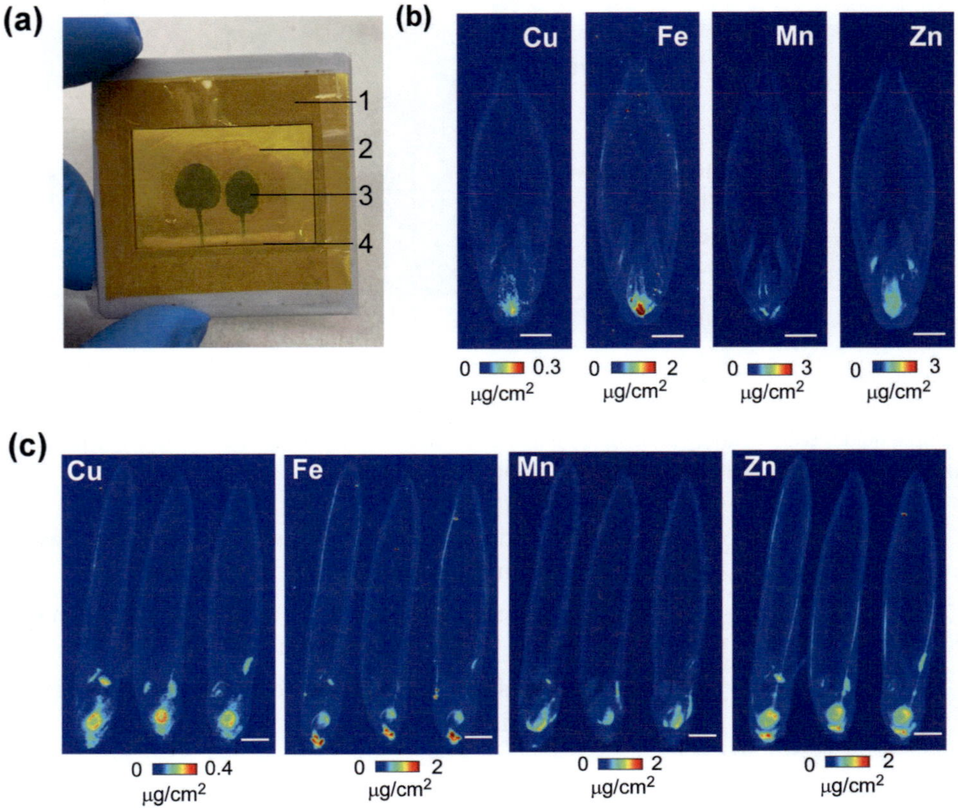

Fig. 3 The application of wet chambers in analyzing fresh plant tissues. (**a**) An example of the wet chamber. Leaves of 20-day-old *Arabidopsis thaliana* were detached and mounted in a 1 × 1.5 inches film frame. 1, the outer layer of Kapton film. 2 and 3, the leaves mounted between Kapton tape and Kapton film. 4. A string of wet Kimwipe that keeps the humidity of the chamber. (**b** and **c**) The elemental maps of a fresh wheat flower (**b**) and a fresh *Brachypodium distachyon* flower (**c**). Samples were detached and mounted in wet chambers in the same setup shown in (**a**). Bars = 1 mm

1. Choose the plant tissues: make sure they are long enough to reach the sample station once they are mounted. If flowers are being analyzed, make sure they are dissected properly prior to sample mounting.
2. Attach the samples to a piece of Kapton tape. Then place the samples along with the Kapton tape on the sample holder (*see* **Note 4**).

3.1.2 Seeds and Grains

Uneven surfaces of seeds and grains may cause issues in 2D-SXRF experiments. Longitudinal sections cut with a cryotome (if feasible) can prevent this problem [12]. However, 2D-SXRF analysis of intact seeds and grains can still be informative for small plant species (e.g., *Arabidopsis thaliana* and *Brachypodium distachyon*). Polishing the seeds/grains can also help to create a relatively flat surface for 2D-SXRF.

Example 1: Mounting Unhusked *Brachypodium distachyon* Grains

1. Place a piece of Kapton tape on a sample holder.
2. Rinse the grains with 95% ethanol to wash away contaminants.
3. Arrange the grains on the tape with the embryos facing in the same direction.

Example 2: Preparation of Partially Polished Beans

1. Open a dry bean and keep the side attached to the embryo.
2. Use a series of sandpapers (from low to high grit) to polish the bean until the embryo is close to the surface (Fig. 4; *see* **Note 5**).
3. Rinse the polished beans with 95% ethanol three times.
4. Follow the procedure for mounting unhusked *B. distachyon* grains to mount the partially polished beans.

3.1.3 Fixed Sample Preparation

1. Fix the tissues in 2% glutaraldehyde in 0.05 M cacodylate buffer pH 7.4 for 2 h on ice in a vacuumed environment.
2. Rinse the tissues in 0.05 M cacodylate buffer pH 7.4 for 10 min.
3. Repeat step 2 for two more times.
4. Dehydrate the tissues in an ethanol series: 25%, 50%, 70%, 95%, and 100%. Incubate the samples in each solution for 10 min. The final two 100% ethanol changes are done with a molecular sieve, which traps any water that may be in the absolute ethanol.
5. Transfer the tissues into absolute acetone which has a molecular sieve for 10 min.
6. Repeat step 5 for one more time (*see* **Note 6**).
7. Embedding in embedding media: This process takes at least 2 days. The first step is a 1 part epoxy to 3 parts acetone for 4–8 h, followed by 1:1 for 4–8 h; 3:1 for 8 h, and 100% epoxy for 12 h.
8. Polymerize the embedded tissues in molds at 60 °C for 12 h.
9. Cut the embedded samples until the area of interest is close to one surface. The proximate dimension is 1–10 mm in diameter with a thickness of 0.1–0.5 mm.

3.2 Scanning

3.2.1 Conventional 2D SXRF Scanning

A conventional 2D scanning SXRF experiment requires an X-ray source, most likely with some kind of focusing optic; a sample placed at 45° relative to the incident beam, with the surface at the focal distance from the optic; and an energy-dispersive X-ray detector placed at 90° relative to the X-ray beam, most likely alongside a digital signal processor for increased speed (Fig. 1). The beam

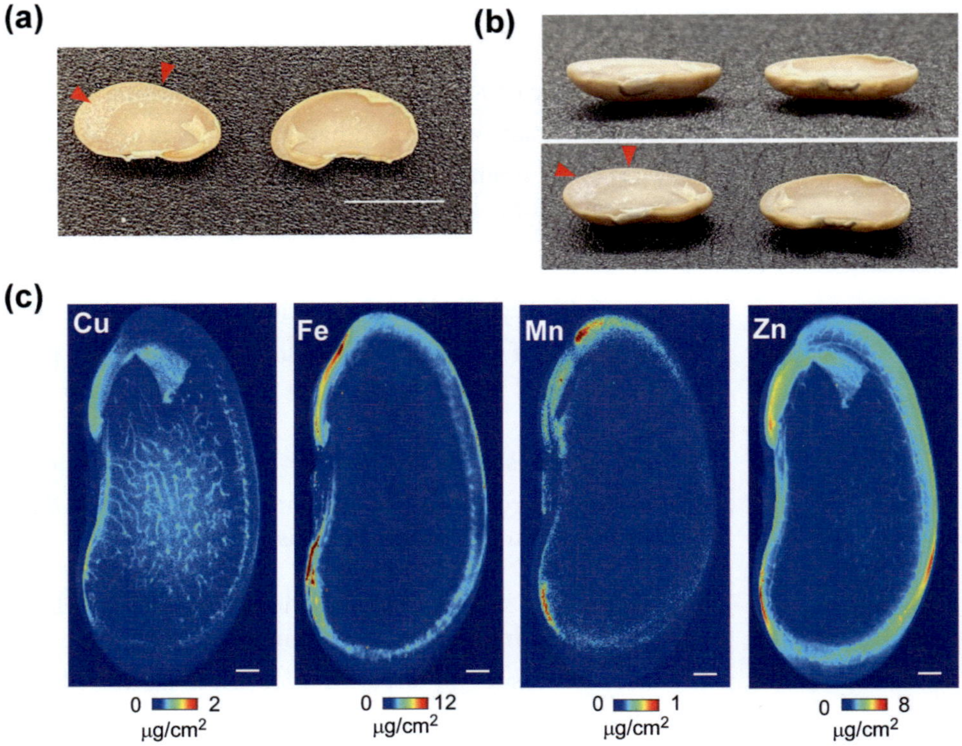

Fig. 4 SXRF images of a partially polished bean. (**a**) Top view of partially polished bean (left) and unpolished bean. (**b**) Side view of partially polished bean (left) and unpolished bean. Red arrows depict the polished area of each bean half. (**c**) SXRF images showing the distribution of copper, iron, manganese and zinc in a polished bean half

energy determines which elements are detected by exciting atoms whose ionization energy is below the beam energy (https://xdb.lbl.gov Table 1.1). Emitted X-rays (XRF signal) have characteristic energies for each element (https://xdb.lbl.gov Table 1.2).

Optimizing for Low Concentrations

A frequent challenge is looking for SXRF signals from elements at low concentrations. One approach to increase measured signal above a background of noise would be to increase the dwell time; however, signal-to-noise is proportional to the square root of the number of measurements. In other words, a 2× increase in signal-to-noise requires measuring 4× as long, but a 10× increase in signal-to-noise requires measuring 100× as long. For low signals, the diminishing returns on signal-to-noise enhancement with increased dwell time are not always practical or feasible. An alternative or additional approach is to take advantage of the fact that an element's ionization cross-section has a maximum just above its ionization energy (https://xdb.lbl.gov Table 1.5). With a tunable X-ray source, such as a synchrotron beamline, the incident X-ray energy can be selected slightly above the ionization energy for the element of interest by 1 or a few keV to maximize the probability of generating XRF signal from that element.

Example

Copper in *A. thaliana* exists at very low concentrations, with wild-type levels below 0.1 μg/cm^2 [11]. The copper ionization energy/absorption edge is 8.979 keV. The incident energy was lowered from the beamline's more typical 17 keV to12.8 keV in order for copper to more efficiently absorb X-rays and generate fluorescence. The energy was not shifted closer to the copper absorption edge in order to avoid the coincidence of Compton scattering background and the SXRF peaks of interest.

Quantification

An important advantage of SXRF is that it can be a quantitative method, because the peak area of the measured signal for each element is proportional to the concentration of that element. The expression of the relationship between measured counts and concentrations can vary slightly depending on the analytical software used (e.g., PyMCA, Praxes, and GeoPIXE are all used in various applications at CHESS), but in general, the relationship has the form:

$$N_i = C_i{}^* \, I_0{}^* \, M_i.$$

where N_i represents the measured number of photons in the peaks for a specific element (from fitting), C_i represents the concentration of the element, I_0 represents the incident flux (photons/second), and M_i is a term taking into account the incident beam energy, the detector solid angle and efficiency, fluorescence yield of the specific element, and attenuation effects due to the sample composition and any air or other filters between the sample and detector.

Therefore, to extract concentrations, it is critical during data collection to:

1. Measure the incident flux I_0 at every pixel in each XRF scan by placing an ion chamber in the beam path before X-rays reach the sample (usually upstream of the focusing optic).
2. Determine the portions of M_i that are set by the detector geometry and efficiency:
   (a) Perform an SXRF scan of a calibration standard of known concentration (e.g., Micromatter XRF calibration standards), keeping sample-detector geometry and distances identical to the setup for the samples of interest.
   (b) Provide an approximate definition for the sample matrix, including the major chemical components and thickness as a proxy for the more complex real sample.
   (c) For Praxes XRF analysis software:

(i) Load and fit the data for the calibration foil, setting the beam energy, and sample matrix/thickness for the foil, and setting the monitor efficiency parameter to 1.

(ii) Note the average concentration of the major element in the foil. Calculate the new monitor efficiency: monitor efficiency = (known concentration of calibration foil [μg/cm^2])/(measured concentration for monitor efficiency of 1 [μg/cm^2]).

(iii) Enter the new value for the monitor efficiency in Praxes and re-run the fit. The resulting concentration should match the reported value for the foil.

(iv) Use this new value for the monitor efficiency when fitting sample data (changing the matrix composition and thickness to describe the sample).

3.2.2 Confocal/3D Scanning SXRF

The following procedures are for the collection of calibration and measurement data:

1. Alignment and calibration.
   (a) Mount AXO thin-film standard at the sample position, and align the film such that the 3D probe volume of the confocal SXRF system intersects with the film surface.
   (b) Perform a scan in the direction normal to the surface, collecting fluorescence data at each point.
   (c) For each element in the thin film standard, extract the peak area of one or more XRF peaks from that element. Typically this is performed with analysis software such as PyXRF, PyMCA. The peak area vs. scan position will have the shape of a normal distribution.
   (d) Using the known values for each element in the standard of the fluorescence cross-section σ_F, the areal mass density σ_i, and the incident intensity Φ_0, fit the data from each peak of the scan of the thin film to the equation,

$$\Phi_i(x) = \Phi_0 \sigma_F \sigma_i \frac{\widetilde{\eta}}{\sigma_x \sqrt{2\pi}} \exp\left(-x^2/2\sigma_x^2\right),$$

in order to extract the integral sensitivity, $\eta(E_i)$, from each curve, and for each element of interest i. For elements not present in the calibration film, the functional form of $\eta(E)$ on XRF energy must be interpolated from known peaks. The functional form of $\eta(E)$ for each Energy E at which elements of interest produce X-ray fluorescence constitutes calibration of the instrument. Once $\eta(E)$ is known, the XRF peak area $\Phi(x)$ from a particular element in a sample whose elemental concentrations are unknown may be converted to volumetric concentration $\rho_i(x)$ as described below.

2. Sample mounting.
   (a) C-SXRF with CCAs may be conducted on either fixed or unfixed samples. Fixed samples are advantageous since they permit the cutting of a flat face near the incident-beam region (see Fig. 2a), which eases the alignment of the region of interest to the confocal volume.
   (b) Fixed, cut samples were attached to a Kapton film stretched across a window-frame-like sample holder (Fig. 2c). The sample holder permits the mounting of many samples, all of which be scanned in sequence.
   (c) The sample frame is placed onto the sample holder (Fig. 2b). Using the optical microscope incorporating a cross-hair locating the position of the incident beam, the scanning region of interest for each sample in the sample holder is identified in the units of the sample stage motors. For this step, the CCA is retracted from its scan position close to the sample to prevent the CCA from obscuring the view of the sample surface.
   (d) After the scan regions are identified and recorded, the thin-film reference sample is briefly mounted in place of the sample and the CCA is brought back to the scan position, permitting the CXRF alignment to be checked and, if necessary, refined. Misalignments of as much as 0.002 mm can affect the scan resolution, making this step important.
   (e) Remount the sample in place of the thin-film reference.
   (f) For each sample on which a map is desired, a single, 1D scan is performed to locate the sample surface, so that the absorption coefficients ($\mu_{i,\ \mathrm{lin}}$) and scan depth (x) are known precisely. Next, a 2D map is conducted at the desired depth – typically 0.04–0.1 mm below the surface. (see Fig. 5a).
3. Data Analysis.
   (a) Use PyXRF (or other software with a similar function) to extract XRF peak areas for each position in the scan.
   (b) Using the calibrated integral sensitivity function $\eta(E)$ determined above, along with the known scan depths, use the equation

$$\Phi_i(x) = \Phi_0 \sigma_F \widetilde{\eta} \overline{\rho_i(x)} e^{-\mu_{i,\mathrm{lin}} x}$$

   to convert peak areas to elemental volumetric density for each element of interest. Maps obtained with this procedure of *A. Thaliana* petioles and stems, in addition to optical micrographs of the samples measured, are shown in Fig. 5.

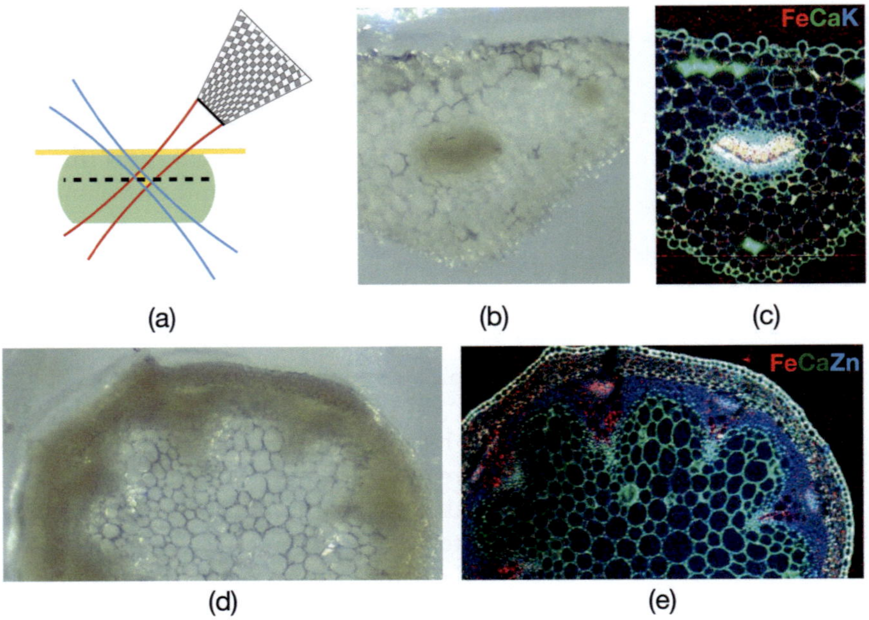

Fig. 5 C-SXRF images of plant tissues. (**a**) Schematic representation of the geometry for C-SXRF measurements, indicating the nominal path of the incident (blue) and fluorescence (red) X-ray beams, and the measurement plan for X-ray maps (dashed line), approximately 0.06 mm below the sample surface. (**b**) optical micrograph and (**c**) combined RGB elemental map of iron, calcium, and potassium obtained from a C-SXRF scan of an *A. thaliana* petiole. (**d**) optical micrograph and (**e**) RGB elemental map of iron, calcium, and zinc obtained from a C-SXRF scan of a *A. thaliana* stem

4 Notes

1. Thick Kapton film and Kapton tape may increase the background noise.
2. A wide range of sample holders is available according to the setup at each beamline. Three-D printed sample holders can be customized to meet the needs.
3. Fresh plant tissues in a wet chamber could tolerate 6–7 h of scans.
4. Sealing the whole sample with Kapton tape helps to prevent fast dehydration during a scan. However, if the concentration of the elements in the tissues that need to be analyzed (e.g., copper) is low, it is better not to cover the area of interest with Kapton tape.
5. It is noteworthy that sandpaper may increase element contamination (e.g., iron and silicon) in polished samples. Thus, it is essential to rinse the samples with 95% ethanol after polishing.
6. The reason to use acetone is that the epoxy is miscible with acetone, not ethanol.

Acknowledgments

The work in the OKV lab is supported by NSF-IOS Awards #1656321 and #1754966; USDA-NIFA Awards # 2021-67013-33798 and #2018-67013-27418. We thank Dr. Raymond P. Glahn and Dr. Jason A. Wiesinger for providing bean samples for SXRF analysis. The work by LS and ARW, and the development of high-resolution C-SXRF, was conducted at the Cornell High Energy Synchrotron Source, which is supported by the National Science Foundation and NIH-NIGMS under NSF-DMR-0225180. The SRX Beamline of the National Synchrotron Light Source II, a U.S. Department of Energy (DOE) Office of Science User Facility, operated for the DOE Office of Science by Brookhaven National Laboratory under Contract No. DE-SC0012704. We thank Andrew Kiss for his invaluable assistance in the use of SRX.

References

1. Kopittke PM, Punshon T, Paterson DJ, Tappero RV, Wang P, Blamey FPC, van der Ent A, Lombi E (2018) Synchrotron-based X-ray fluorescence microscopy as a technique for imaging of elements in plants. Plant Physiol 178:507–523
2. Kopittke PM, Lombi E, van der Ent A, Wang P, Laird JS, Moore KL, Persson DP, Husted S (2020) Methods to visualize elements in plants. Plant Physiol 182:1869–1882
3. Ackerman CM, Lee S, Chang CJ (2017) Analytical methods for imaging metals in biology: from transition metal metabolism to transition metal signaling. Anal Chem 89:22–41
4. Punshon T, Ricachenevsky FK, Hindt MN, Socha AL, Zuber H (2013) Methodological approaches for using synchrotron X-ray fluorescence (SXRF) imaging as a tool in ionomics: examples from *Arabidopsis thaliana*. Metallomics 5:1133–1145
5. Punshon T, Guerinot ML, Lanzirotti A (2009) Using synchrotron X-ray fluorescence microprobes in the study of metal homeostasis in plants. Ann Bot 103:665–672
6. McRae R, Bagchi P, Sumalekshmy S, Fahrni CJ (2009) In Situ imaging of metals in cells and tissues. Chem Rev 109:4780–4827
7. Zhai Z, Gayomba SR, Jung HI, Vimalakumari NK, Piñeros M, Craft E, Rutzke MA, Danku J, Lahner B, Punshon T, Guerinot ML, Salt DE, Kochian LV, Vatamaniuk OK (2014) OPT3 is a phloem-specific iron transporter that is essential for systemic iron signaling and redistribution of iron and cadmium in Arabidopsis. Plant Cell 26:2249–2264
8. Yan J, Chia JC, Sheng H, Jung HI, Zavodna TO, Zhang L, Huang R, Jiao C, Craft EJ, Fei Z, Kochian LV, and Vatamaniuk OK (2017) Arabidopsis pollen fertility requires the transcription factors CITF1 and SPL7 that regulate copper delivery to anthers and jasmonic acid synthesis. Plant Cell 29: 3012–3029
9. Yamaguchi N, Ishikawa S, Abe T, Baba K, Arao T, Terada Y (2012) Role of the node in controlling traffic of cadmium, zinc, and manganese in rice. J Exp Bot 63:2729–2737
10. Sheng H, Jiang Y, Rahmati M, Chia JC, Dokuchayeva T, Kavulych Y, Zavodna TO, Mendoza PN, Huang R, Smieshka LM, Miller J, Woll AR, Terek OI, Romanyuk ND, Piñeros M, Zhou Y, Vatamaniuk OK (2021) YSL3-mediated copper distribution is required for fertility, seed size and protein accumulation in Brachypodium. Plant Physiol 186:655–676
11. Chia JC, Yan J, Ishka MR, Faulkner M, Huang R, Smieska L, Woll A, Tappero R, Jiao C, Fei Z, Piñeros M, Kochian LV, Vatamaniuk OK (2023) Loss of OPT3 function decreases phloem copper levels and impairs crosstalk between copper and iron homeostasis and shoot-to-root signaling in Arabidopsis thaliana. Plant Cell https://doi.org/10.1093/plcell/koad053
12. Li Y, Dhankher OP, Carreira L, Lee D, Chen A, Schroeder JI, Balish RS, Meagher RB (2004) Overexpression of phytochelatin synthase in *Arabidopsis* leads to enhanced arsenic tolerance and cadmium hypersensitivity. Plant Cell Physiol 45:1787–1797

Chapter 15

A Simple Semi-hydroponic System for Studying Iron Homeostasis in Maize

Stavroula Fili and Elsbeth Walker

Abstract

Hydroponic-based systems for plant growth allow control of the nutrients that plants take up through the provided nutrient solution. Different formulations of nutrient solutions enable the study of nutrient deficiencies in plants. Here we describe a procedure for setting up a simple semi-hydroponic system to grow maize seedlings. The system can be set up on a small or large scale, depending on the number of individuals studied. A modified nutrient solution is used for growing maize seedlings in iron-replete and iron-depleted conditions. This setup allows for studies of iron-deficiency responses in maize.

Key words Hydroponicsiron, Maize, Gene expression, Iron nutrition, *Zea mays*

1 Introduction

Hydroponics is a method of growing plants without soil using a water solution. The solution contains a defined nutrient formulation capable of supporting healthy plant growth. Hydroponic systems are widely used in plant biology research to study plant responses to biotic [1–3] and abiotic stresses [4–6]. Because plants grow in a nutrient solution without soil, these systems are ideal for studying roots, which can be accessed easily and isolated intact [7, 8]. Additionally, since the nutrient solution composition can be easily controlled, hydroponic systems are a preferred method for studying nutrient deficiencies and nutrient homeostatic processes in plants [9–11].

Different hydroponic systems and variations have been developed over the years. However, they all rely on delivering the nutrients to the plant by exposing the plant root system to the solution. Roots can be fully submerged in the solution (classic hydroponics), fully exposed to the air, in which case they are sprayed with the solution at specific time intervals (aeroponics), or there can be many variations between these two.

Jeeyon Jeong (ed.), *Plant Iron Homeostasis: Methods and Protocols*, Methods in Molecular Biology, vol. 2665,
https://doi.org/10.1007/978-1-0716-3183-6_15,

In hydroponics, plants are typically grown on some type of inert medium, which provides some root anchorage and general support to the plant. These media can be either natural (such as wood fiber, clay pellets, coconut, or rice husks) or artificial, such as polystyrene peanuts. Some media are made from organic materials or are rich in minerals and may not be suitable for experimental systems in which close control of the solution composition is essential.

The final component that is required in all hydroponic systems is aeration [12]. Roots need to be well supplied with oxygen to prevent hypoxia. Root aeration can be achieved by using air pumps designed for use in aquariums for classic hydroponics or be achieved by using wet and dry cycles for other types of systems.

Here we present a custom semi-hydroponic system for growing maize seedlings. The system uses polypropylene buckets as nutrient solution containers and mesh pots for growing the seedlings with expanded clay pebbles as an inert medium. The nutrient solution is provided to the roots using a water pump that creates a fountain that sprays the roots. The roots are initially not submerged into the solution but become partially submerged as the plants grow. We have used this system to study iron homeostasis in maize by obtaining iron-regulated gene expression data. We have repeatedly evaluated the capability of this system to reflect the plants' physiological response between replete and depleted iron conditions by using genes that serve as markers for iron deficiency.

2 Materials

To study iron homeostasis in maize, plants are grown in a modified Hoagland's nutrient solution. The pH of the nutrient solution is maintained using MES as a buffering agent, which is non-toxic to plants [13]. All stock solutions are prepared using distilled water. The final nutrient solution is prepared using reverse-osmosis (RO) type water.

2.1 Stock Solutions

Prepare each of the following stock solutions using reagents from your preferred supplier. The suggested concentrations of the stocks are given in parentheses to make calculations easier for the final nutrient solution.

1. Stock solutions for macronutrients: Prepare each solution in a volume of 500 mL and sterilize by autoclaving (*see* **Note 1**):
   (a) 1 M Potassium phosphate monobasic (KH_2PO_4; 1000× stock)
   (b) 1 M Calcium nitrate ($Ca(NO_3)_2$; 1000× stock)

   (c) 1 M Potassium nitrate (KNO_3; 1000× stock)
   (d) 1 M Magnesium sulfate ($MgSO_4$; 1000× stock)
2. Stock solutions for the micronutrients: Prepare each solution in a volume of 200 mL and sterilize by autoclaving (*see* **Note 1**):
   (a) 250 mM Calcium chloride ($CaCl_2$; 10,000× stock)
   (b) 250 mM Boric acid, 10,000× (H_3BO_3; 10,000× stock)
   (c) 200 mM Manganese sulfate ($MnSO_4$; 100,000× stock)
   (d) 200 mM Zinc sulfate ($ZnSO_4$; 100,000× stock)
   (e) 50 mM Copper sulfate ($CuSO_4$; 100,000× stock)
   (f) 50 mM Molybdic acid (H_2MoO_4; 100,000× stock).
3. 0.5 M MES buffer: Adjust the pH of the MES buffer to 5.9 with 1 N KOH. Next, sterilize the MES buffer by filtration using a 0.2 μm filter unit (*see* **Note 2**).
4. Iron(II) Sulfate/EDTA Chelated Solution (0.01 M stock, 100×) (*see* **Note 3**).

2.2 Nutrient Solution

Depending on the total volume needed, the nutrient solution can be prepared in glass bottles (1 L, 2 L) or polypropylene carboys with a dispenser for bigger volumes. The amounts of stock solutions per 1 L of nutrient solution are given in Table 1. These amounts are multiplied accordingly for the total volume of solution needed for a specific experiment.

2.3 Semi-hydroponic Equipment

The semi-hydroponic system can be used on a small or large scale, depending on the number of individuals to be studied. In the small-scale system, up to 6 individuals can be grown per container, which gives enough space for multiple biological replicates. In the large-scale system, up to 18 individuals can be grown per container (up to three individuals per mesh pot). The large-scale system is ideal if a segregating population is grown or if more than one genotype needs to be grown in the same container. You will need to use a drill with the suggested hole saw diameters below to create holes in the lid for placing the mesh pots (Fig. 1a). Up to six holes can be created with these sizes around the perimeter of the lid (Fig. 1b). A smaller hole must be created for the water pump cable to pass through (Fig. 1b). Also, make a cut between the small hole and one of the adjacent bigger holes (Fig. 1b).

As a support medium, we use expanded clay pebbles as they do not affect the pH and lack nutrients. Wash the pebbles thoroughly before use. Sterilize them by washing with 10% bleach and rinsing thoroughly or autoclaving in a dry cycle.

2.3.1 Small-Scale System Equipment

1. High-density polyethylene (HDPE) buckets of 1-gallon capacity with snap-on lids, black color

Table 1
Stock and final nutrient concentrations used for the hydroponic solution

Nutrient stock	Stock concentration (M)	Final concentration (mM)	Volume of stock for 1 L nutrient solution (mL)
KH_2PO_4 (1000×)	1	1	1
$Ca(NO_3)_2$ (1000×)	1	1	1
KNO_3 (1000×)	1	1	1
$MgSO_4$ (1000×)	1	1	1
$CaCl_2$ (10,000×)	0.25	0.025	0.1
H_3BO_3 (10,000×)	0.25	0.025	0.1
$MnSO_4$ (100,000×)	0.2	0.002	0.1
$ZnSO_4$ (100,000×)	0.2	0.002	0.1
$CuSO_4$ (100,000×)	0.05	0.0005	0.01
H_2MoO_4 (100,000×)	0.05	0.0005	0.01
MES-KOH	0.5	2	4
$FeSO_4$-EDTA (100×)	0.001	0.1	10

2. Two-inch slotted mesh pots with thin lip rims, preferably black color
3. Two-inch diameter hole saw drill bit
4. Mini submersible water pump capable of creating a flow rate of up to 95GPH
5. Plug-in mechanical timer capable of creating 30-min intervals

2.3.2 Large-Scale System Equipment

1. High-density polyethylene (HDPE) buckets of 3.5-gallon capacity with snap-on lids, black color
2. Three-inch slotted mesh pots with thin lip rims, preferably black color
3. Three-inch diameter hole saw drill bit
4. Submersible water pump capable of creating a flow rate of up to 250GPH
5. Plug-in mechanical timer capable of creating 30-min intervals

Fig. 1 Equipment used to create a hydroponic container. Here, the small-scale system is shown as an example. A hole saw drill bit attachment is used to create the holes for the mesh pots (**a**). A smaller hole must be created using the regular drill bit, and a cut must be made between the smaller hole and one of the adjacent big holes (**b**). The pump has a small pipe attachment on the top that creates the fountain (**c**). The pump is placed in the middle of the bucket (**d**). The cable is passed through the cut and into the small hole (**e**)

3 Methods

3.1 Disinfecting and Setting Up the Semi-hydroponic System

All equipment parts must be disinfected before setting up a new experiment.

1. For the disinfecting solution, prepare enough volume of 10% bleach for the number and size of buckets you are using. For example, 1 L of the solution is adequate for a small-size bucket, while 3 L is adequate for a big-size bucket.
2. Fill the buckets with the solution and wipe all surfaces of the buckets, including the lids.
3. The mesh pots can be dipped into the disinfecting solution inside the buckets.
4. Place the pumps into the solution and fix them in the middle of the bucket. Then, plug them in so they can run with the disinfecting solution.
5. The disinfecting solution must stay in contact with the equipment for at least 10 min. After this time, rinse all parts thoroughly with water. Run the pump with water for a few minutes to ensure there is no bleach residue inside the pump.
6. Set up the system by placing the water pump in the middle of the bucket (Fig. 1d).
7. Pass the pump cable through the opening and place the lid on the bucket (Fig. 1e).
8. Fill the mesh pots halfway with clean, sterilized clay beads.
9. Place the mesh pots in the lid openings.

3.2 Sterilizing and Germinating Maize Seeds

Sterilize the maize seeds in 50 mL disposable polypropylene tubes. Use up to 30 seeds per tube.

1. Prepare the maize sterilization solution (15% bleach, 0.01% Triton-X 100) for a total volume of 40 mL for each tube by mixing 6 mL bleach with 34 mL sterilized ultra-pure water and adding 4 μL of Triton-X 100.
2. Place the seeds into the solution, close the tubes tightly, and place them in an orbital shaker. Let the tubes shake for 15 min at 120 RPM.
3. Pour out the sterilization solution and wash the seeds with many rounds of sterilized ultra-pure water, shaking them by hand and pouring out the water. Repeat at least 5 times, or until the water does not foam from leftover Triton-X 100.
4. Add one last volume of water and leave the seeds in for 5 min before planting.
5. Plant the seeds directly in the mesh pots by placing them on the clay beads.
6. Add the appropriate volume of RO water in each bucket (1 L for the small system; 5 L for the large system) (*see* **Note 4**). Cover the buckets with aluminum foil.
7. Run the system covered at 24 °C for 3 to 5 days to allow germination (*see* **Note 5**). We suggest replenishing the water daily to prevent contamination.
8. After coleoptiles have emerged, gently fill the rest of the mesh pot with clay pebbles.

3.3 Preparing and Replenishing the Nutrient Solution

The nutrient solution should be provided to plants when the second leaf starts to emerge (V1 stage). Full-strength nutrient solution can be damaging to young seedlings, so the first nutrient solution supplied to the plants is a half-strength solution (1/2×). After 3 days a 75% strength (0.75×) can be used, and at the third medium change, full strength (1×) solution is begun. If browning of the leaf tips occurs, the nutrient solution strength should stay reduced for additional time until the plants adapt or grow larger.

1. Calculate the amounts of nutrients needed for your total volume, according to Table 1 (*see* **Note 6**). Use 1 L of hydroponic solution for each 1-gallon bucket and 5 L of hydroponic solution for each 5-gallon bucket.
2. Using a graduated cylinder (500 mL for small scale, 2 L for large scale), measure RO water to about 80% of its capacity (about 400 mL or 1.6 L, correspondingly) (see Table 2). Pipette the macronutrients into the water (*see* **Note 7**). Bring the volume to 100% with water. Add this amount to the hydroponic solution container (bottle or carboy). Keep track of the

Table 2
Step-by-step procedure to formulate the hydroponic solution. The appropriate volumes are given, depending on the required total volume of hydroponic solution

Total volume needed	1 L	2 L	5 L	10 L
Graduated cylinder size to use	250 mL	500 mL	1 L	2 L
Step 1: Measure RO water	200 mL	400 mL	800 mL	1.6 L
Step 2: Add appropriate volumes of macronutrients				
Step 3: Bring volume to	250 mL	500 mL	1 L	2 L
Step 4: Add to hydroponic solution container				
Step 5: Measure RO water	200 mL	400 mL	800 mL	1.6 L
Step 6: Add appropriate volumes of micronutrients				
Step 7: Bring volume to	250 mL	500 mL	1 L	2 L
Step 8: Add to hydroponic solution container				
Volume in the container so far	500 mL	1 L	2 L	4 L
Step 9: Measure RO water	200 mL	400 mL	800 mL	1.6 L
Step 10: Add MES buffer and iron stock, if making "+Fe" solution Add MES buffer only, if making "−Fe" solution				
Step 11: Bring volume to	250 mL	500 mL	1 L	2 L
Volume in the container so far	750 mL	1.5 L	3 L	6 L
Step 12: Measure RO water	50 mL	100 mL	1 L	2 L
Step 13: Add to hydroponic solution container				
Volume in the container so far	800 mL	1.6 L	4 L	8 L
Step 14: Adjust pH to 5.6 with 1 N KOH				
Step 15: Measure RO water	200 mL	400 mL	1 L	2 L
Step 16: Add to hydroponic solution container to bring to total volume				

volume added to the hydroponic solution container with each addition.

3. Repeat the previous step with the micronutrients this time and add this amount to the hydroponic solution container.
4. Repeat the previous step by adding the MES buffer and iron stock (*see* **Note 3**) this time and adding this amount to the hydroponic solution container.
5. Fill in the remaining amount to 80% of the total volume (see Table 2).
6. Adjust the pH to 5.6 using 1.0 N KOH (*see* **Note 8**). Bring the hydroponic solution to total volume with water.

7. Add the appropriate volume of hydroponic solution in each bucket (1 L for the small system; 5 L for the large system).
8. Place the buckets in a growth chamber operating in a 16 h/8 h light/dark cycle at 24 °C.
9. Run the system by plugging the pumps through the mechanical timer. Program the timer according to the manufacturer's instructions to operate in a 1.5 h/0.5 h on/off cycle.

The hydroponic nutrient solution must be replaced every 3 days (*see* **Note 9**).

3.4 Shifting Plants to Different Iron Conditions

To induce iron deficiency conditions, two formulations of hydroponic solution must be prepared; one solution will have the full composition ("+Fe") while the other will lack the iron stock ("−Fe") (Fig. 2).

Seedlings growing in an iron-replete solution for 9–12 days (typically at the early V3 stage) can be transferred to a nutrient solution lacking iron for up to 3 days. At the three-day point, chlorosis is usually evident, and molecular markers of iron deficiency have been activated.

1. Prepare the "+Fe" nutrient solution for the control plants as described in the previous section.
2. Prepare the "−Fe" nutrient solution in the same way but omit the iron stock.
3. Before shifting plants to the "−Fe" nutrient solution, the roots and the clay pebbles must be thoroughly rinsed. The "+Fe"

Fig. 2 Maize seedlings grown in the small-scale semi-hydroponic system captured at late V3 stage. Interveinal chlorosis is visible in seedlings that have been switched to "−Fe" solution for 3 days (bucket on the right)

control plants must also be handled in the same way as the −Fe plants, to ensure that changes in gene expression caused by handling and solution change are uniform in all samples. Pour out the previous solution from the buckets, lift the lid with the plants, and wash the roots and clay pebbles with running water. Rinse the top of the lid, the inside of the bucket, and the pump.

4. Add the fresh "+Fe" and "−Fe" nutrient solutions into the separate buckets.

3.5 Collecting Tissues and Performing Gene Expression Measurements

Root and shoot tissue can be collected from plants growing in "+Fe" and "−Fe" conditions to measure gene expression. Tissues used for RNA extraction must be flash-frozen in liquid nitrogen immediately after collection. If not processed after collection, tissues can be frozen in a −80 °C freezer for longer preservation.

Collecting root tissues from plants growing in the semi-hydroponic system is quick and easy, as the medium (clay pebbles) can be easily removed from the roots.

1. Lift the individual mesh pots containing the plants and empty the clay pebbles into a disposing container to clean the roots from the medium.
2. Blot the roots with Kimwipes to remove excess liquid before freezing.
3. Place the roots into 50 mL disposable polypropylene tubes and freeze them in liquid nitrogen (*see* **Note 10**).
4. Grind the roots into a fine powder using RNase-free pre-frozen mortar and pestle (*see* **Note 11**) using liquid nitrogen liberally to maintain the tissues in a deeply frozen state. Avoid allowing any thawing, as this is very detrimental to RNA isolation.
5. Use the amount of ground tissue needed for your preferred method of RNA extraction.
6. Use qRT-PCR to measure the expression of iron-regulated genes. The genes *ZmYs1* (GRMZM2G156599) and *ZmIRO2* (GRMZM2G057413) can be used as reliable markers of iron deficiency (Fig. 3).

4 Notes

1. The stock solutions for macro and micronutrients are prepared individually to discourage the precipitation of salts that may occur when various salts are mixed at high concentrations.
2. Store the MES buffer and the iron stock solution (wrapped with aluminum foil to prevent light degradation of EDTA) at 4 °C. All other stock solutions can be kept at room temperature.

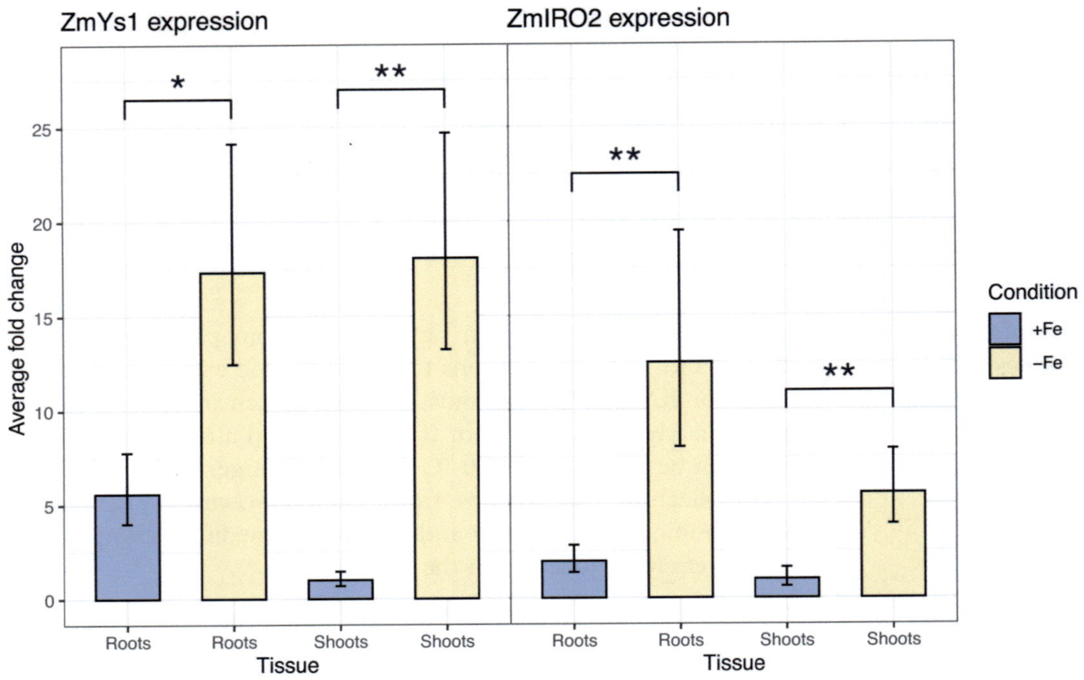

Fig. 3 Iron-regulated gene expression in maize wild-type seedlings growing in "+Fe" and "−Fe" conditions. The genes ZmYs1 and ZmIRO2 are upregulated during iron deficiency in roots and shoots. Three biological replicates were used per sample type

3. As the iron source, we suggest using Iron (II) Sulfate/EDTA Chelated Solution (0.01 M stock, 100×) (PhytoTech, F318).
4. Never run the pumps without adding the water or the hydroponic solution first. Submersible water pumps will overheat and break down when running without liquid.
5. Make sure that the pebbles are getting fully wet by adjusting the flow rate of the pump to move the solution to a height that appropriately wets the pebbles.
6. The total volume would be typically 2 L for the small-scale system and 10 L for the large-scale system. See Table 2 for an explanation of the volumes added.
7. Adding water first into the cylinder prevents later precipitation of the stock solutions when they are mixed together. Concentrated stocks will precipitate when mixed directly.
8. If adjusting the pH for a large scale, make sure to stir the carboy very well after adding KOH. After stirring, use a small beaker to take a sample of the solution to measure the pH.
9. If salt build-up is observed after running the system with nutrient solution, the excess salts must be removed from the system. This can be performed by flushing the system, i.e., running the system only with water for a few minutes, two to

three times. Flushing can be used as a prevention method for salt build-up every 3 days, before refreshing the nutrient solution.

10. To measure iron-regulated gene expression, we collect whole root systems using the above process, but individual root types can be easily dissected if needed.
11. Mortars and pestles are covered in aluminum foil and baked at 200 °C for at least 4 h to destroy RNase molecules. After they cool down, they are placed at −20 °C the night before use.

References

1. Sijmons PC et al (1991) Arabidopsis thaliana as a new model host for plant-parasitic nematodes. Plant J 1(2):245–254
2. Djonovic S et al (2007) A proteinaceous elicitor Sm1 from the beneficial fungus Trichoderma virens is required for induced systemic resistance in maize. Plant Physiol 145(3): 875–889
3. Luna E et al (2011) Callose deposition: a multifaceted plant defense response. Mol Plant-Microbe Interact 24(2):183–193
4. Metwally A et al (2003) Salicylic acid alleviates the cadmium toxicity in barley seedlings. Plant Physiol 132(1):272–281
5. Ehlert C et al (2009) Aquaporin-mediated reduction in maize root hydraulic conductivity impacts cell turgor and leaf elongation even without changing transpiration. Plant Physiol 150(2):1093–1104
6. Arfan M, Athar HR, Ashraf M (2007) Does exogenous application of salicylic acid through the rooting medium modulate growth and photosynthetic capacity in two differently adapted spring wheat cultivars under salt stress? J Plant Physiol 164(6):685–694
7. Umehara M et al (2008) Inhibition of shoot branching by new terpenoid plant hormones. Nature 455(7210):195–U29
8. Chen YL et al (2011) Development of a novel semi-hydroponic phenotyping system for studying root architecture. Funct Plant Biol 38(5):355–363
9. Assuncao AGL et al (2010) Arabidopsis thaliana transcription factors bZIP19 and bZIP23 regulate the adaptation to zinc deficiency. Proc Natl Acad Sci U S A 107(22):10296–10301
10. Qin L et al (2012) The high-affinity phosphate transporter GmPT5 regulates phosphate transport to nodules and nodulation in soybean. Plant Physiol 159(4):1634–1643
11. Nguyen NT, McInturf SA, Mendoza-Cozatl DG (2016) Hydroponics: a versatile system to study nutrient allocation and plant responses to nutrient availability and exposure to toxic elements. J Vis Exp 113
12. Smeets K et al (2008) Critical evaluation and statistical validation of a hydroponic culture system for Arabidopsis thaliana. Plant Physiol Biochem 46(2):212–218
13. Bugbee BG, Salisbury FB (1985) An evaluation of mes (2(n-morpholino)ethanesulfonic acid) and amberlite irc-50 as ph buffers for nutrient solution studies. J Plant Nutr 8(7):567–583

Chapter 16

Optimizing Fe Nutrition for Algal Growth

Anne G. Glaesener and Sabeeha S. Merchant

Abstract

Chlamydomonas is an excellent reference system for dissecting the impact of iron (Fe) nutrition on photosynthetic and other metabolisms. The operational definition of four stages of Fe nutrition is described and a guide to the practical use of these stages is offered, specifically the preparation of media and growth of mixotrophic cultures. A key consideration is the impact of carbon metabolism on the expression of Fe-containing enzymes and hence the Fe quota. The absolute concentration of Fe in the medium is less determinative of gene expression than the Fe available on a per-cell basis. In nature, algal cells may transition from Fe-replete to -deficient to -limited during a bloom.

Key words *Chlamydomonas reinhardtii*, Iron, Elemental profile, Photoheterotrophic growth

1 Introduction

While research on iron (Fe) nutrition in plants has largely focused on Fe-uptake pathways and their regulation, photosynthetic microbes such as the unicellular green alga *Chlamydomonas reinhardtii* provide excellent experimental systems for understanding Fe metabolism at the cellular and subcellular level, including the relation of Fe nutrition to photosynthesis and acclimation to abiotic stress, among other biological processes. The convenience of culturing a microorganism in a defined medium allows quantitative analysis of Fe nutrition responses by itself and in the context of macro- or micro-nutrient deficiencies. Several patterns in Fe homeostasis have been established in this reference organism, including photosystem remodeling and preferential retention of some pathways and key Fe-dependent proteins in response to suboptimal Fe supply [1–3]. In the era of elaborate and complex experimental designs and large-scale comparisons of experimental datasets, including various -omics studies, having comparable and reproducible starting material is key [4].

Numerous molecular, genetic, and genomic resources make *Chlamydomonas reinhardtii* an excellent reference organism for

Jeeyon Jeong (ed.), *Plant Iron Homeostasis: Methods and Protocols*, Methods in Molecular Biology, vol. 2665,
https://doi.org/10.1007/978-1-0716-3183-6_16,

studying diverse biological processes [5]. Its metabolic versatility makes it attractive to study cellular responses to external stimuli. Its photosynthetic apparatus is very similar to that of land plants [6–8]. As a fast-growing single-cell organism, it responds quickly to environmental perturbation. Three basic techniques are generally used to generate poor Fe nutrition in the laboratory. The first is to limit the available Fe in the medium by chelators [9–11]. The second technique is to limit intracellular Fe by using an Fe-transport mutant [11, 12]. In either situation, Fe assimilation is hindered, but not completely blocked. Cells do acquire Fe, but more slowly, in the latter case through lower-affinity, less-specific transporters. The third and preferable approach is to control the amount of Fe added to the medium, which is described in more detail in this protocol. The use of alkaline pH to reduce Fe solubility is not generally applied to Chlamydomonas because of the harm high pH does to Chlamydomonas cells [4] and the relative ease of omitting or tightly controlling the amount of available Fe in the growth medium.

Chlamydomonas is cultured in a simple, defined medium where the concentration of metal ions can be selectively controlled. Because of high metabolic demand for Fe, Fe depletion in such defined media is relatively easy to accomplish. However, several precautions should be followed to ensure that the amount of Fe in the medium is consistent and that the experiments are therefore reproducible (i.e., the only Fe present is that which is consciously added). The study of Fe homeostasis in Chlamydomonas is routinely performed in the context of four graded Fe nutrition stages: excess, replete, deficient, and limited [2, 3, 13, 14]. These states were defined by the evaluation of phenotype and Fe-responsive gene expression in response to controlled medium Fe content. Specifically, components of the Fe-uptake pathway are used as sentinel genes for Fe status, as well as physiological characteristics and the intracellular Fe quota (Fig. 1).

One intention of modern-day experiments, especially comparative and large-scale experiments, is the gain of knowledge, beyond the individual treatment. Multi-variate comparisons with temporal components have inherent layers of variability, some of them systematic [4]. A goal for good data analysis is to extract sensible data from the background. Growth conditions can vary dramatically and small changes that remain unnoticed and unrecorded can cause extensive background variability. Thus, the goal is to generate highly reproducible cell cultures with consistent physiological Fe status as starting material for a wide range of experimental analyses.

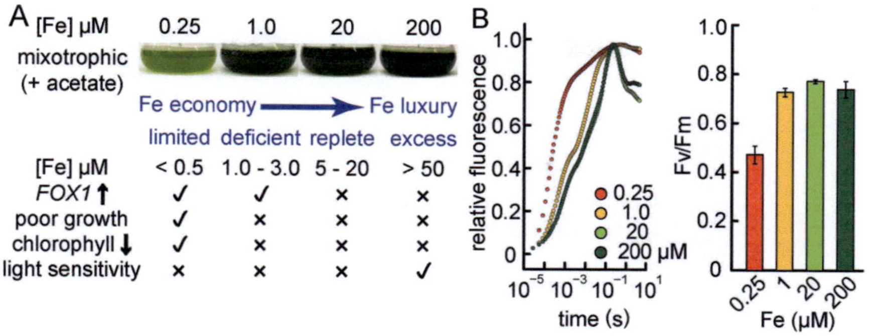

Fig. 1 Distinct stages of Fe nutrition in Chlamydomonas. (**a**) Representative cultures of Chlamydomonas at the defined Fe nutritional states, limited, deficient, replete, and excess. The different physiological states are distinguished based on sentinel gene expression and phenotype. (**b**) Fluorescent transients corrected for background and normalized to 1 for comparison (left) and Photosystem II efficiency (F_v/F_m, right). The mean of three replicates is shown, and error bars represent SD

2 Materials

Prepare all solutions using ultrapure water (purifying deionized water further to reach conductivity of 18 MΩ cm^{-1} at 20 °C, Milli-Q-purified or alike) and ultra-pure reagents. Prepare and store reagents at room temperature unless specifically indicated otherwise. Diligently follow all safety precautions and regulations, including proper personal protective equipment (PPE), the use of chemical fume hoods when handling hazardous chemicals, and follow appropriate waste disposal regulations.

2.1 Acid Washing of Glass- and Plasticware

1. Ultrapure hydrochloric acid, ~18.5%, 6 M H^+.
2. Ultrapure water.

2.2 Prepare Trace Metal Grade Stock Solutions

Prepare stock solutions for TAP growth medium [15, 16] using high-purity reagents (*see* **Note 1**), acid-washed graduated cylinders (*see* **Note 2**), ultrapure water, and store solutions in acid-washed plastic bottles (*see* **Notes 2** and **3**). To dissolve salts, add them to water in a graduated cylinder and cover the cylinder with parafilm and carefully invert the closed cylinder several times. An acid-washed stir bar may also be used but carries a larger contamination risk.

1. Tris-Acetate, 100× (1 L, 2 M Tris, 1.7 M Acetate)

Add about 500 mL ultrapure water to a 1 L graduated cylinder. Weigh 242 g Tris base and transfer to cylinder. Add 100 mL concentrated glacial acetic acid. Make up to 1 L with ultrapure water and mix well. Store at 4 °C for up to 6 months. Use 10 mL stock solution to prepare 1 L of TAP medium.

2. Phosphate Solution, 100× (1 L, 0.567 mM K_2HPO_4, 0.375 mM KH_2PO_4)

Add about 900 mL ultrapure water to a 1 L graduated cylinder. Weigh 11.95 g K_2HPO_4 (anhydrous) and 6.05 g KH_2PO_4 (anhydrous), transfer to cylinder, and dissolve. Make up to 1 L with ultrapure water and mix well. Store at 4 °C for up to 6 months. Use 10 mL stock solution to prepare 1 L of TAP medium.

3. Beijerink's Solution, 100× (1 L, 7.5 mM NH_4Cl, 0.34 mM $CaCl_2$, 0.4 mM $MgSO_4$)

Add about 300 mL ultrapure water to a 500 mL graduated cylinder. Weigh 5 g $CaCl_2 \cdot 2\ H_2O$ and transfer to cylinder. Add about 500 mL ultrapure water to a 1 L graduated cylinder. Weigh 40 g NH_4Cl and 10 g $MgSO_4 \cdot 7\ H_2O$ and transfer both to the 1 L cylinder. Mix well, then combine the two solutions. Make up to 1 L with ultrapure water and mix well. Store at 4 °C for up to 6 months. Use 10 mL stock solution to prepare 1 L of TAP medium.

4. Pre-EDTA (125 mM EDTA, pH 8.0)

Add about 250 mL ultrapure water to a 500 mL graduated cylinder. Weigh 13.96 g $Na_2EDTA \cdot 2\ H_2O$, add to cylinder, and dissolve. Titrate to pH 8.0 with trace element grade KOH (~1.7 g), make up to 300 mL with ultrapure water and mix well. Store at 4 °C for up to 12 months.

5. Pre-Molybdenum (285 μM Mo)

Add about 200 mL ultrapure water to a 250 mL graduated cylinder. Weigh 0.088 g $(NH_4)_6Mo_7O_{24} \cdot 4\ H_2O$, add to cylinder, dissolve, make up to 250 mL with ultrapure water, and mix well. Store at 4 °C for up to 12 months. This solution is only used to prepare the Molybdenum stock solution.

6. Pre-Selenium (1 mM Se)

Add about 200 mL ultrapure water to a 250 mL graduated cylinder. Weigh 0.043 g Na_2SeO_3, add to cylinder, dissolve, make up to 250 mL with ultrapure water, and mix well. Store at 4 °C for up to 12 months. This solution is only used to prepare the Selenium stock solution.

7. EDTA, 1000× (25 mM EDTA)

Mix 200 mL ultrapure water with 50 mL Pre-EDTA (Subheading 2.2, solution 4). Store at 4 °C for up to 12 months. Use 1 mL stock solution to prepare 1 L of TAP medium.

8. Molybdenum, 1000× (28.5 μM Mo)

Mix 225 mL ultrapure water with 25 mL Pre-Molybdenum (Subheading 2.2, solution 5). Store at 4 °C for up to 12 months. Use 1 mL stock solution to prepare 1 L of TAP medium.

9. Selenium, 1000× (0.1 mM Se)

Mix 225 mL ultrapure water with 25 mL Pre-Selenium (Subheading 2.2, solution 6). Store at 4 °C for up to 12 months. Use 1 mL stock solution to prepare 1 L of TAP medium.

10. Zinc, 1000× (2.5 mM Zn, 2.75 mM EDTA)

Add about 200 mL ultrapure water to a 250 mL graduated cylinder. Weigh 0.18 g $ZnSO_4 \cdot 7\ H_2O$, add to cylinder, dissolve, and add 5.5 mL Pre-EDTA solution (Subheading 2.2, solution 4). Make up to 250 mL with ultrapure water and mix well. Store at 4 °C for up to 12 months. Use 1 mL stock solution to prepare 1 L of TAP medium.

11. Manganese, 1000× (6 mM Mn, 2.75 mM EDTA)

Add about 200 mL ultrapure water to a 250 mL graduated cylinder. Weigh 0.297 g $MnCl_2 \cdot 4\ H_2O$, add to cylinder, dissolve, and add 12 mL Pre-EDTA solution (Subheading 2.2, solution 4). Make up to 250 mL with ultrapure water and mix well. Store at 4 °C for up to 12 months. Use 1 mL stock solution to prepare 1 L of TAP medium.

12. Iron, 1000× (20 mM Fe, 22 mM EDTA)

Add about 200 mL ultrapure water to a 250 mL graduated cylinder. Weigh 0.58 g Na_2CO_3 and 2.05 g Na_2EDTA (do not use Pre-EDTA solution), add to cylinder, and dissolve. Add 1.35 g $FeCl_3 \cdot 6\ H_2O$ after the first two components are dissolved. Make up to 250 mL with ultrapure water and mix well. Store at 4 °C for up to 12 months. Use 1 mL stock solution to prepare 1 L of TAP medium.

13. Copper, 1000× (2 mM cu, 2 mM EDTA)

Add about 200 mL ultrapure water to a 250 mL graduated cylinder. Weigh 0. $CuCl_2 \cdot 2\ H_2O$, add to cylinder, dissolve, and add 4 mL Pre-EDTA solution (Subheading 2.2, solution 4). Make up to 250 mL with ultrapure water and mix well. Store at 4 °C for up to 12 months. Use 1 mL stock solution to prepare 1 L of TAP medium.

2.3 Prepare TAP Growth Medium

1. TAP Medium, 1x (1 L, Replete)

Add about 800 mL ultrapure water to an acid-washed 1 L graduated cylinder. Add 10 ml Tris-Acetate stock (Subheading

2.2, solution 1), 10 mL Phosphate solution (Subheading 2.2, solution 2), 10 mL Beijerinck's solution (Subheading 2.2, solution 3), 1 ml each of the seven trace element stocks (Subheading 2.2, solutions 7–13). Make up to 1 L with ultrapure water and mix well. Aliquot into acid-washed Erlenmeyer flasks, most commonly 100 mL into 250 mL flasks. Autoclave for 20 min, 121 °C and let cool to room temperature before use.

2. Fe-Free TAP Medium, 1x (1 L)

Add about 800 mL ultrapure water to an acid-washed 1 L graduated cylinder. Add 10 ml Tris-Acetate stock (Subheading 2.2, solution 1), 10 mL Phosphate solution (Subheading 2.2, solution 2), 10 mL Beijerinck's solution (Subheading 2.2, solution 3), 1 ml each of the EDTA, Molybdenum, Selenium, Zinc, Manganese and Copper trace element stocks (Subheading 2.2, solutions 7–11, 13), omit the Fe trace element stock (Subheading 2.2, solution 12). Make up to 1 L with ultrapure water and mix well. Aliquot into acid-washed Erlenmeyer flasks, most commonly 100 mL into 250 mL flasks. Add Fe trace element stock (Subheading 2.2, solution 12) to individual flasks as needed (*see* Methods for more details). Autoclave for 20 min, 121 °C and let cool to room temperature before use. Use within 2 days of preparation.

3 Methods

Carry out all procedures at room temperature unless specifically instructed otherwise. Avoid metal contamination at each step (*see* **Note 3**).

3.1 Generate Starter Cultures

1. Prepare 40 mL TAP growth medium as described in Subheading 2.3, solution 1. Mix well and add to a 100 mL Erlenmeyer flask. Autoclave for 20 min, 121 °C and let cool to room temperature before use.
2. Inoculate with a small, but visible lump of Chlamydomonas cells (~1 × 1 mm, or depending on strain) from a one- to two-week grown TAP-agar plate (*see* **Notes 4** and **5**).
3. Grow the culture at 24 °C with agitation (180 rpm or what is required by the particular strain used) in constant illumination with a light intensity of 100–200 μmol photons·m^{-2}·s^{-1} (*see* **Note 6**) for 4–5 days or until a cell density of ~5 × 10^6 cells·ml^{-1} is reached.
4. Prepare 300 mL Fe-free TAP growth medium as described in Subheading 2.3, solution 2. Mix well and aliquot 100 mL each into a 250 mL Erlenmeyer flask. Add 15 μL Fe stock (Subheading 2.2, solution 12) to each 100 mL Fe-free TAP to create 3 μM TAP for Fe-reduced precultures (*see* **Note 7**). Autoclave

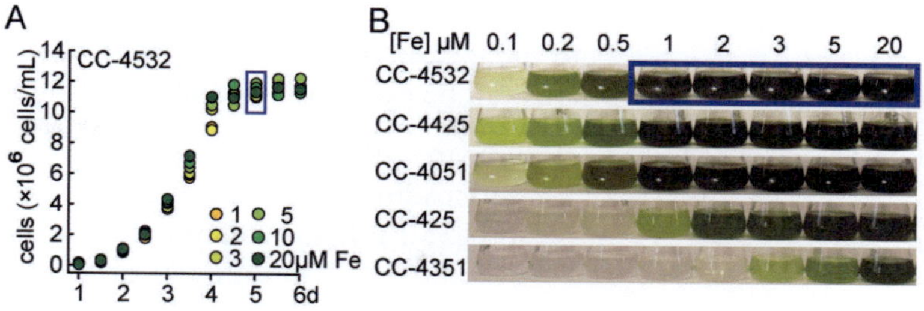

Fig. 2 Growth of Chlamydomonas. Cells were grown photoheterotrophically in a wide range of extracellular Fe. Representative cultures are shown. (**a**) Cell densities of strain CC-4532 in Fe-deficient and Fe-replete conditions, ranging from 1 to 20 μM. By the late-logarithmic growth phase (day 4), the Fe-deficient cultures lag slightly behind the replete cultures, indicative of poor Fe nutrition, thus 3 μM medium is used to generate Fe-pool depleted pre-cultures. The flasks equivalent to 5d after inoculation are boxed in panel B. (**b**) Growth of 5 wild type strains in a wide range of Fe conditions. Strains CC-425 and CC-4351 lack a fully functional cell wall (*see* **Note 9**). Cultures were photographed after 5 days of growth

for 20 min, 121 °C and let cool to room temperature before use.

5. Inoculate from the mid-logarithmic culture by transferring 100 μL cells from the 40 mL culture into each of the three 3 μM Fe TAP flasks.
6. Grow the culture at 24 °C with agitation in constant illumination with a light intensity of 100–200 μmol photons·m^{-2}·s^{-1} (*see* **Note 6**) for 4–5 days or until a cell density of ~8 × 10^{6} cells·ml^{-1} is reached (*see* **Note 8** and Fig. 2).

   Use this late-logarithmic culture to inoculate cultures for various Fe states.

3.2 Acid Washing of Glass- and Plasticware

1. In a fume hood, place the glass- and plasticware needed, including graduated cylinders, plastic bottles, or glass flasks, in a secondary container.
2. Pour clean 6 M hydrochloric acid into the vessels; fill above the intended fill line. Cover open containers with plastic wrap or similar and let soak for 1 h (range of 15 min up to overnight). Remove acid and repeat with fresh acid, let soak for 1 h. Remove acid and rinse each item at least six times with ultrapure water to remove acid.
3. Acid-washed items should be used within 1 week of preparation. Keep items covered with plastic wrap to avoid contamination with dust carrying metal contaminations.

3.3 Experimental Growth of Chlamydomonas Cells in all Four Fe States

In this section, all four Fe-stages, as discussed in the introduction, are prepared, and grown in three independent replicate cultures. This can be easily adjusted to the experimental needs.

1. Prepare 1200 mL Fe-free TAP growth medium as described in Subheading 2.3, solution 2. Mix well and aliquot 100 mL each into 250 mL Erlenmeyer flasks. Add 1.25 μL Fe stock (Subheading 2.2, solution 12) to three flasks to create Fe-limited TAP (0.25 μM Fe), add 15 μL Fe stock to three flasks to create Fe-deficient TAP (1 μM Fe), add 100 μL Fe stock to three flasks to create Fe-replete TAP (20 μM Fe), and add 1 mL Fe stock to three flasks to create Fe-excess TAP (200 μM Fe). Autoclave for 20 min, 121 °C and let cool to room temperature before use (*see* **Notes 10** and **11**).
2. Inoculate each of the twelve flasks with 50 μL cells from the late-logarithmic culture containing 3 μM Fe.
3. Grow the cultures at 24 °C with agitation in constant illumination with a light intensity of 100–200 μmol photons·m^{-2}·s^{-1} (*see* **Note 6**) until mid-logarithmic growth phase is reached.
4. Collect and analyze cells based on the experimental design (*see* **Notes 12** and **13**, Figs. 1 and 3).

4 Notes

1. Ultra-high-purity chemicals are used to make Fe-free stock solutions, which are stored in acid-washed, metal-free plasticware. Certificates of analysis specifying the trace metal composition are generally available before the purchase of a particular lot of the stock chemicals and can be used to estimate the amount of contaminating metals in the prepared medium. The lot number should be recorded as part of the metadata. These chemicals should be kept separate from other laboratory chemicals to avoid accidental metal contamination.
2. The use of clean glassware and plasticware is of paramount importance. All culture flasks and reusable plasticware are rinsed twice with 6 N HCl to displace metal ions and rinsed at least six times with ultrapure water to remove the acid. Metal is bound to layers within the glass. The acid-washing removes only surface-bound ions, therefore, glassware should be used promptly after acid-washing.
3. At all times, effort should be made to avoid contamination with metals (wear gloves, no metal spatulas, and protection from dust). Ideally, Fe-deficient media prepared in acid-washed glass flasks should be used immediately. It is recommended that the media not be stored for more than a couple of days.

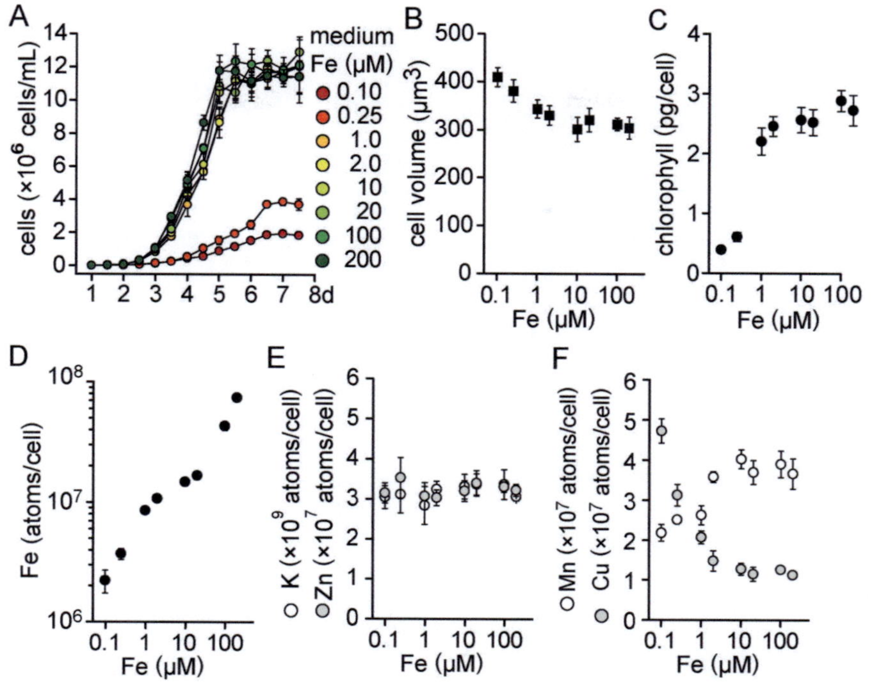

Fig. 3 Characteristics of photoheterotrophic Chlamydomonas wild type strain CC-4532 at the four defined stages of Fe nutrition. Two extracellular concentrations per stage are shown, 0.1 and 0.25 μM for limited, 1 and 2 μM for deficient, 10 and 20 μM for replete and 100 and 200 μM for excess. The mean of three to five replicates is shown, error bars represent SD. (**a**) Growth, measured by a hemocytometer and expressed as $\times10^6$/mL. (**b**) Cell volume, estimated from cell diameter, measured with an image-based automated cell counter. (**c**) Chlorophyll content, measured spectrophotometrically after extraction of pigments into acetone: methanol. (**d**) Cellular Fe content, (**e**) cellular K and Zn content and (**f**) cellular Mn and Cu content measured by ICP-MS/MS (*see* **Note 13**)

4. Chlamydomonas strains are best kept on Fe-replete TAP-agar plates for maintenance and long-term storage. The preparation of solid Fe-deficient media, if desired for experimental purposes, requires washing of the agar with EDTA to remove contaminating metal ions [17].
5. A larger volume of growth medium, for example 100 mL, can be inoculated directly from an agar plate. This may require a larger number of cells from the agar plate, or the cell culture will take longer to reach late logarithmic growth phase to be used in subsequent liquid cultures.
6. Use a mixture of cool white and warm white light [15] at a ratio of 2:1 of 4100 K cool white and 3000 K warm white fluorescence bulbs or LEDs with comparable spectrum. Determine light intensity and spectral characteristics in several spots within the incubator to expose experimental cultures to comparable photon qualities and quantities (Fig. 4). Record the quantity

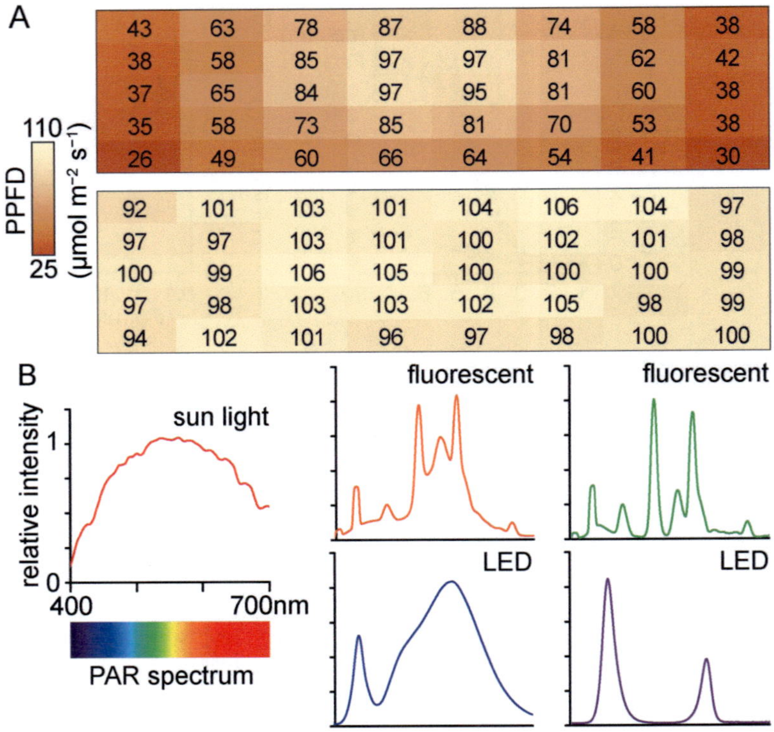

Fig. 4 Light intensity and spectral characteristics. (**a**) Light intensity landscape inside a growth chamber with fluorescent light bulbs (top) and LEDs (bottom). (**b**) Representative light spectra of five different light sources, normalized to 1 for comparison and limited to the PAR spectral range (400–700 nm). Sunlight (red), cool white and warm white fluorescent bulbs (orange), white fluorescent bulbs (green), broad band, full spectrum LED (blue) and narrow band, targeted spectrum LED (purple)

and quality of light as associated metadata, including the method used by the instrument. Most digital light meters have the option to measure PAR, the photosynthetically active radiation, and PPFD, the photosynthetic photon flux density, between the wavelengths of 400 and 700 nm, which is the portion of the light spectrum utilized for photosynthesis.

7. Chlamydomonas cultures grown through a full growth cycle (inoculation to stationary growth phase) consume ~3 μM Fe from the growth medium. Use reduced-Fe precultures to decrease internal Fe pools.

8. Cultures inoculated from plate will take slightly longer to reach mid-logarithmic growth phase due to a lag in doubling, whereas liquid cultures inoculated from a mid- to late-logarithmic liquid culture will reach mid-logarithmic growth phase faster.

9. Most Chlamydomonas wildtype strains show the same distinct stages, although some strains, most significantly strains lacking a fully functional cell wall and newly generated mutants, need to be phenotyped for their specific Fe concentration thresholds corresponding to the physiological Fe stages. FEA proteins, associated within the periplasm in a cell wall-containing strain, assist in Fe assimilation, without a fully functioning cell wall, those proteins are lost to the growth medium [18].
10. All growth vessels should be treated the same, no matter if they are going to contain Fe-excess or Fe-limited growth medium to reduce variation within experimental sets of cultures. All twelve flasks used in Subheading 3.3.1. are acid-washed and the same ultra-high-purity solutions are used to generate the growth medium.
11. Cell density of experimental cultures has an impact on the amount of Fe actually available to the cells. The concentration (μM) noted for the growth medium at each of the stages is at the time of inoculation. As cells grow and divide, they use up some or all the Fe available in the growth medium, creating different amounts of Fe available in the medium as function of time and thus potentially different physiological states, at distinct positions in the cellular growth dynamic (Fig. 5). In other words, cells inoculated at 3 μM Fe in the medium will transition from replete to deficient to limited as the culture transitions from lag to log to stationary growth phase.
12. It should be noted that the defined Fe nutrition stages above are described for cells grown photoheterotrophically (light and acetate). The same stages can be described for phototrophic cells (light and CO_2), but the exact Fe concentration thresholds are distinct, as is the cellular response to those concentrations. These stages are also distinct for cells grown purely heterotrophically (dark and acetate).
13. Monitoring whether deficiency in one metal affects the intracellular concentration of other metal ions is important as well. For instance, non-selective transporters may be induced, which will inadvertently bring in multiple metal ions. The cell may purposefully change the concentration of other metals as seen for Fe and manganese and for zinc and copper levels in Chlamydomonas [18, 19]. Therefore, it is imperative to measure all intracellular metal contents during Fe nutrition experiments or monitor additional markers of other micronutrient deficiencies (Fig. 6, also see Fig. 3e, f).

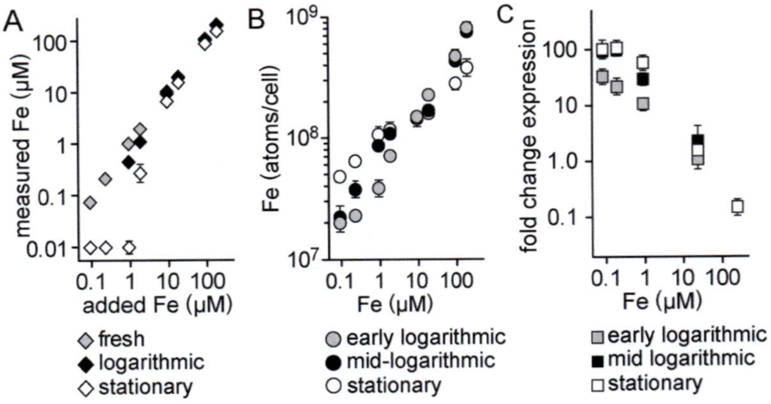

Fig. 5 Fe status during the growth phases. (**a**) Fe content in the spent medium, at inoculation (grey), in mid-logarithmic (black) and stationary growth phase (white). (**b**) Intracellular Fe as a function of extracellular Fe, in early logarithmic (grey), mid-logarithmic (black), and stationary (white) growth phase. (**c**) Relative expression of *FOX1*, a sentinel gene for the cellular Fe status, determined by qPCR as a function of cell growth in early logarithmic (grey), mid-logarithmic (black), and stationary growth phase (white). The mean of three to five replicates is shown, error bars represent SD

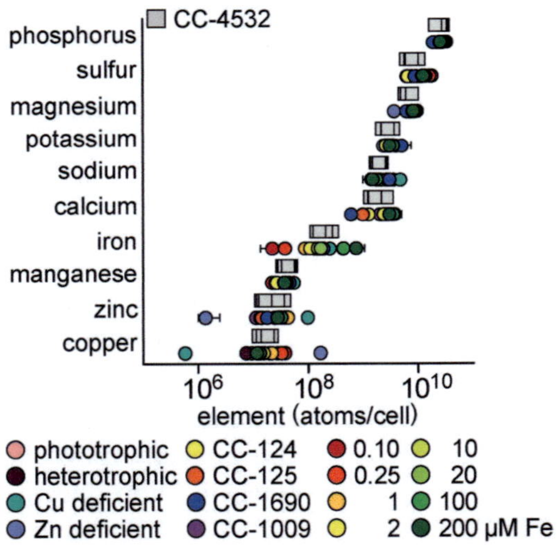

Fig. 6 Elemental profile of Chlamydomonas cells. Elements per cell in one wild type strain grown under replete photoheterotrophic conditions (grey boxes), compared to various conditions, some of which change the elemental profile of the cultures. The mean of three to five (colored) or 26 (grey) replicates is shown, error bars represent SD

Acknowledgments

This work was supported by Department of Energy (DOE) grant DE-SC0020627.

References

1. Naumann B, Busch A, Allmer J et al (2007) Comparative quantitative proteomics to investigate the remodeling of bioenergetic pathways under iron deficiency in *Chlamydomonas reinhardtii*. Proteomics 7(21):3964–3979
2. Terauchi AM, Peers G, Kobayashi MC et al (2010) Trophic status of *Chlamydomonas reinhardtii* influences the impact of iron deficiency on photosynthesis. Photosynth Res 105:39–49
3. Urzica EI, Casero D, Yamasaki H et al (2012) Systems and trans-system level analysis identifies conserved iron deficiency responses in the plant lineage. Plant Cell 24(10):3921–3948
4. Hui C, Schmollinger S, Strenkert D et al (2022) Simple steps to enable reproducibility: culture conditions affecting Chlamydomonas growth and elemental composition. Plant J 111(4):995–1014
5. Salomé PA, Merchant SS (2019) A series of fortunate events: introducing Chlamydomonas as a reference organism. Plant Cell 31(8): 1682–1707
6. Morrissey J, Guerinot ML (2009) Iron uptake and transport in plants: the good, the bad, and the ionome. Chem Rev 109(10):4553–4567
7. Kobayashi T, Nishizawa NK (2012) Iron uptake, translocation, and regulation in higher plants. Annu Rev Plant Biol 63:131–152
8. Thomine S, Vert G (2013) Iron transport in plants: better be safe than sorry. Curr Opin Plant Biol 16(3):322–327
9. Weger HG, Espie GS (2000) Ferric reduction by iron-limited *Chlamydomonas* cells interacts with both photosynthesis and respiration. Planta 210:775–781
10. Rubinelli P, Siripornadulsil S, Gao-Rubinelli F, Sayre RT (2002) Cadmium- and iron-stress-inducible gene expression in the green alga *Chlamydomonas reinhardtii*: evidence for H43 protein function in iron assimilation. Planta 215:1–13
11. Terzulli A, Kosman DJ (2010) Analysis of the high-affinity iron uptake system at the *Chlamydomonas reinhardtii* plasma membrane. Eukaryot Cell 9:815–826
12. Dix DR, Bridgham JT, Broderius MA et al (1994) The *FET4* gene encodes the low affinity Fe(II) transport protein of *Saccharomyces cerevisiae*. J Biol Chem 269:26092–26099
13. Moseley JL, Allinger T, Herzog S et al (2002) Adaptation to Fe-deficiency requires remodeling of the photosynthetic apparatus. EMBO J 21:6709–6720
14. Long JC, Merchant SS (2008) Photo-oxidative stress impacts the expression of genes encoding iron metabolism components in Chlamydomonas. Photochem Photobiol 84:1395–1403
15. Quinn JM, Merchant SS (1998) Copper-responsive gene expression during adaptation to copper deficiency. Methods Enzymol 297: 263–279
16. Kropat J, Hong-Hermesdorf A, Casero D et al (2011) A revised mineral nutrient supplement increases biomass and growth rate in *Chlamydomonas reinhardtii*. Plant J 66:770–780
17. Glaesener AG, Merchant SS, Blaby-Haas CE (2013) Iron economy in *Chlamydomonas reinhardtii*. Front Plant Sci 4:337
18. Allen MD, del Campo JA, Kropat J, Merchant SS (2007) *FEA1*, *FEA2*, and *FRE1*, encoding two homologous secreted proteins and a candidate ferrireductase, are expressed coordinately with *FOX1* and *FTR1* in iron-deficient *Chlamydomonas reinhardtii*. Eukaryot Cell 6: 1841–1852
19. Hong-Hermesdorf A, Miethke M, Gallaher SD et al (2014) Subcellular metal imaging identifies dynamic sites of Cu accumulation in Chlamydomonas. Nat Chem Biol 10(12): 1034–1042

Index

Jeeyon Jeong (ed.), *Plant Iron Homeostasis: Methods and Protocols*, Methods in Molecular Biology, vol. 2665, https://doi.org/10.1007/978-1-0716-3183-6,

MIX
Papier aus verantwortungsvollen Quellen
Paper from responsible sources
FSC® C105338

If you have any concerns about our products,
you can contact us on
ProductSafety@springernature.com

In case Publisher is established outside the EU,
the EU authorized representative is:
Springer Nature Customer Service Center GmbH
Europaplatz 3, 69115 Heidelberg, Germany

Printed by Libri Plureos GmbH
in Hamburg, Germany